Guide to Infectious Diseases by Body System

Jeffrey C. Pommerville, P...
Glendale Community C...

Second Edition

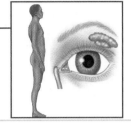
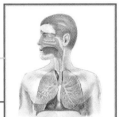
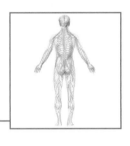
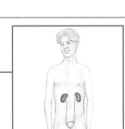
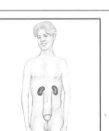
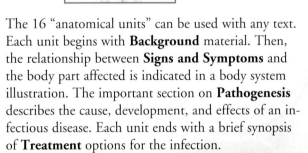

Introduction

For many students entering the allied health and nursing fields, connecting infectious diseases with anatomy is a very necessary and rewarding experience. As I write this second edition, I have seen how important it is for my students to understand how an infectious agent *causes a change from good health* (disease) in the human body. Therefore, I have put together this short *Guide to Infectious Diseases by Body System* to help students understand the relationships between infectious agents and human body systems. Since some diseases affect more than one body system, I have tried to match diseases with the more likely body system affected.

The 16 "anatomical units" can be used with any text. Each unit begins with **Background** material. Then, the relationship between **Signs and Symptoms** and the body part affected is indicated in a body system illustration. The important section on **Pathogenesis** describes the cause, development, and effects of an infectious disease. Each unit ends with a brief synopsis of **Treatment** options for the infection.

I hope you find this guide useful in your microbiology studies and health careers.

Jeffrey Pommerville

1. Bacterial and Fungal Skin Infections

The **skin** is the largest organ of the human body and provides an effective mechanical and chemical barrier to infectious agents. Most skin infections result from scrapes, punctures, or other trauma that break the skin and may go below the **epidermis** and **dermis** to infect the **subcutaneous tissue** (see right). These infections can be caused by the indigenous skin microbiota or by exogenous pathogens.

The entry or infection of pathogens into the skin often results in some form of lesion. These include:

Bullae. Large skin lesions containing liquid

Macules. Flat skin spots that have changed color

Pustules. Raised, red spots on the skin containing pus; without pus they are called **papules**

Scales. Small, flaky pieces of skin

Vesicles. Small, raised, red lesions that contain a clear fluid

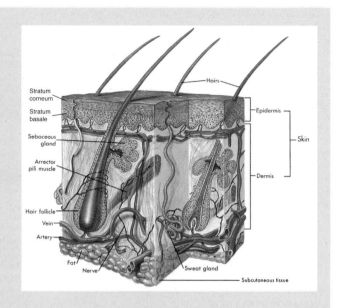

Signs and Symptoms

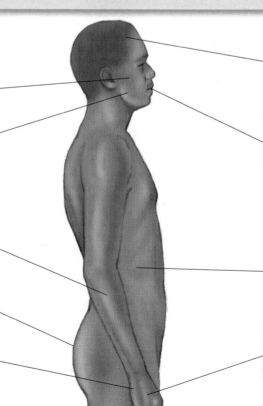

Acne: Presence of whiteheads and blackheads, developing into pustules.

Erysipelas: Itchy, swollen, and reddened rash over face and lower limbs.

Necrotizing fasciitis: High fever, skin reddening, and severe pain with bullae forming at the wound site.

Cutaneous candidiasis: Localized papules and pustules on moist skin folds (breasts, buttocks).

Sporotrichosis: Development of subcutaneous nodules that ulcerate at the puncture site.

Cutaneous anthrax: A painless, reddish-brown papule develops a blister, hardens, and dries to form a black scab (eschar).

Folliculitis: Pustules on scalp, arms, and legs (children); torso and face (adults).

Impetigo: Small macules develop into pustules on face, arms, or legs; ruptures crust over and become yellow-brown scabs.

Dermatophytosis: Pink, itching, blister-like scales (tineas) on the skin.

Scalded skin syndrome: In newborns, bullae form on the skin, followed by peeling of the epidermal layer.

Gas gangrene: Fever, pain, swelling, and blackening of the skin at the wound site.

Pathogenesis

The following is a brief description of several bacterial and fungal infections of the skin.

Bacterial Skin Diseases

Acne: A common, localized, endogenous inflammation of hair follicles on the face and upper torso of adolescents. Sex hormone changes and blocked sebaceous ducts cause rupture of the hair follicle lining, triggering infection by *Propionibacterium acnes*. Severe cases can leave permanent scars on the skin.

Erysipelas: An inflammation of the dermis resulting from the entry of *Streptococcus pyogenes* (group A streptococci) that disseminates as a bright, raised red rash **(Figure 1A)**. The condition can become septicemic and require immediate treatment.

Necrotizing fasciitis: A very severe deep cutaneous infection caused by *S. pyogenes* or *Pseudomonas aeruginosa* that, through a break in the skin, destroys the facial sheets of tissue below the subcutaneous layers of the skin. The infection can spread rapidly, producing an inflammatory response that destroys neurovascular bundles in the fascia, resulting in extensive necrosis (so-called "flesh-eating disease").

Cutaneous anthrax: A serious disease caused by spores of *Bacillus anthracis* that enter through a break in the skin. The initial papule progresses to an ulcerative stage that usually scabs over (black eschar). In some 20% of cases, the infection can become systemic and can be fatal unless antibiotics are used early to prevent dissemination to the blood.

Folliculitis: A superficial infection of hair follicles often caused by *Staphylococcus aureus* or *P. aeruginosa*. Deeper subcutaneous abscesses are called **furuncles** (boils). If neighboring tissues become infected, the enlarged lesion is referred to as a **carbuncle**.

Scalded skin syndrome: A temporary, localized *S. aureus* infection usually restricted to newborns and young children. Release of two bacterial exotoxins causes the epidermis to separate from the dermis, resulting in blistering, reddening, and peeling of the top layers. The now exposed dermal layer looks scalded but heals rapidly. A high fever accompanies the infection.

Impetigo: A relatively common skin lesion that usually occurs in newborns and children from infection by *S. aureus* or *S. pyogenes*. Initial vesicles eventually form pustules that crust over. Although rarely serious, the infection is highly communicable.

Gas gangrene: The disease occurs in wounds and surgical sites that become infected with *Clostridium perfringens*. Systemic effects and heart damage can result from toxin production. Gas forms within muscles or subcutaneous tissue as enzymes destroy the muscle tissue, producing a blackish fluid that oozes from the skin.

Fungal Skin Diseases

Cutaneous candidiasis: An opportunistic, endogenous skin infection usually caused by the overgrowth of *Candida albicans* where excessive moisture exists (e.g., buttocks of infants leads to diaper rash). Systemic infections can occur in immunocompromised hosts.

Dermatophytosis: A common cutaneous infection, caused by several species of *Epidermophyton*, *Microsporum*, and *Trichophyton*, that produces scaling of the skin, loss of hair, or crumbling nails. These "**tineas**" are identified by the infected tissues, such as **tinea pedis**, foot (athlete's foot); **tinea cruris**, groin ("jock itch"); and **tinea capitis**, head (scalp ringworm).

Sporotrichosis: An infection caused by *Sporothrix schenckii* that has contaminated a thorn or sharp vegetation. Small nodules may ulcerate and spread as secondary lesions (**lymphocutaneous sporotrichosis**) along the lymphatics that drain the original lesion **(Figure 1B)**.

✚ Treatment

Treatment of bacterial skin diseases usually is successful using specific antibiotics. Folliculitis is treated through drainage, while acne is treated with topical drying agents. The antifungal drugs miconazole and clotrimazole are used for the tineas and cutaneous candidiasis. Sporotrichosis can be treated with oral itraconazole. Necrotizing fasciitis requires prompt antibiotic therapy and surgical debridement.

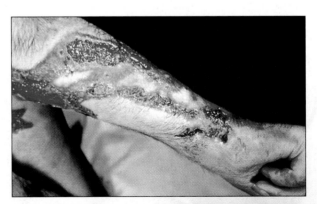

Figure 1A, Erysipelas. This skin lesion involves streptococci infecting the skin and subcutaneous tissue, causing edema and an erythema on the face. (Courtesy of Dr. Thomas F. Sellers, Emory University/CDC)

Figure 1B, Sporotrichosis. The arm of a patient showing the extreme effects of sporotrichosis. What started out as small, painless bumps developed into a lymphocutaneous form. (Courtesy of Dr. Lucille K. Georg/CDC)

2. Viral and Parasitic Skin Infections

Background

There are several viruses and a parasite that cause skin infections and disease in humans, and whose clinical symptoms are commonly diagnosed by their effects on the skin. A few are found solely within the skin, but most skin diseases are systemic, spreading from another body system, such as the blood or respiratory system, before appearing as skin lesions. Several of the viral diseases cause a skin rash, called an **exanthema**. These include chickenpox, measles, and rubella.

Viruses associated with skin diseases belong to one of four groups (based on genome architecture):

Double-Stranded (ds) DNA Viruses

Papovaviridae: Among these viruses are those human papillomaviruses that cause common cutaneous warts.

Poxviridae: This group includes the viruses responsible for smallpox, monkeypox, and molluscum contagiosum.

Herpesviridae: These viruses include the causative agents of cold sores (herpes simplex virus-1; HSV-1); roseola (human herpesvirus-6; HHV-6); and chickenpox and shingles (varicella-zoster virus; VZV).

Single-Stranded (ss) DNA Viruses

Parvoviridae: Within this group is parvovirus B19 that causes erythema infectiosum.

Single-Stranded (ss) RNA Viruses (+ strand)

Togaviridae: This group includes the rubella virus that causes rubella.

Single-Stranded (ss) RNA Viruses (– strand)

Paramyxoviridae: Among these viruses, the rubeola virus is the etiologic agent for measles.

Signs and Symptoms

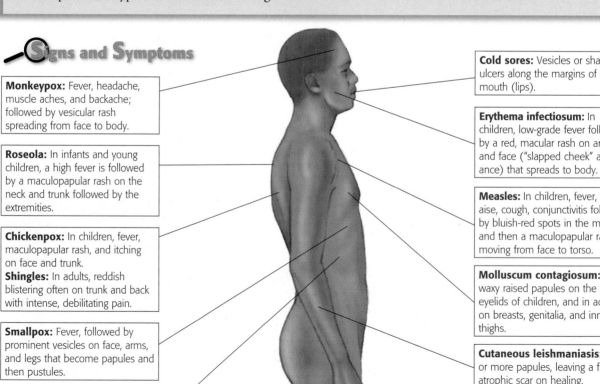

Monkeypox: Fever, headache, muscle aches, and backache; followed by vesicular rash spreading from face to body.

Roseola: In infants and young children, a high fever is followed by a maculopapular rash on the neck and trunk followed by the extremities.

Chickenpox: In children, fever, maculopapular rash, and itching on face and trunk.
Shingles: In adults, reddish blistering often on trunk and back with intense, debilitating pain.

Smallpox: Fever, followed by prominent vesicles on face, arms, and legs that become papules and then pustules.

Rubella: In children, maculopapular red rash and low-grade fever.

Cold sores: Vesicles or shallow ulcers along the margins of the mouth (lips).

Erythema infectiosum: In children, low-grade fever followed by a red, macular rash on arms and face ("slapped cheek" appearance) that spreads to body.

Measles: In children, fever, malaise, cough, conjunctivitis followed by bluish-red spots in the mouth and then a maculopapular rash moving from face to torso.

Molluscum contagiosum: Small, waxy raised papules on the face, eyelids of children, and in adults on breasts, genitalia, and inner thighs.

Cutaneous leishmaniasis: One or more papules, leaving a flat, atrophic scar on healing.

Cutaneous warts: Small, elevated, usually benign skin growths typically on hands and soles of feet

Pathogenesis

The following is a brief description of several virus and parasitic infections of the skin.

Viral Skin Infections

Cold sores (fever blisters): A primary infection with herpes simplex virus-1 (HSV-1) that replicates in epithelial cells on the margins of the lips. T lymphocytes kill the infected cells and the blisters heal as the infection subsides; however, a life-long latent infection is established in the trigeminal ganglia. Later, stress or trauma can reactivate the virus with development of infectious vesicles in the epidermis in the same general area as the primary lesions.

Roseola (exanthema subitum): An acute but benign infection common in infants and young children (reactivation in immunocompromised patients) caused by human herpes viruse 6 (HHV-6). Replicating in the salivary glands, the virus is transmitted through saliva. As the high fever subsides, a pink or rose-colored rash appears on the trunk, neck, and arms.

Chickenpox (varicella): A highly contagious but mild disease of childhood caused by the varicella-zoster virus (VZV). After replication in the respiratory epithelium, VZV is transmitted via the blood to infect skin cells forming vesicles on the trunk and face. The vesicles/pustules rupture and scab over during healing.

Shingles (zoster): Shingles results when VZV (often dormant for decades in sensory ganglia) becomes reactivated and replicates in the peripheral nerves. Burning and extremely painful vesicular lesions typically form on the overlaying skin.

Measles (rubeola): A highly communicable disease caused by the measles virus. After replicating in the respiratory mucosa, small red lesions with white centers (**Koplik spots**) appear in the oral mucosa. The virus spreads via the bloodstream to the skin producing a maculopapular rash that progresses from face to torso and extremities.

Rubella (German measles): A highly contagious but benign disease caused by the rubella virus. The virus typically infects children via respiratory droplets and is spread systemically through the blood, causing a maculopapular rash on the face that spreads to the extremities. Infection of a pregnant woman (first trimester) can lead to congenital defects in the eyes, heart, or brain.

Erythema infectiosum (fifth disease): A contagious but mild disease primarily in children caused by human parvovirus B19. The characteristic rash appears to be the result of immune system activity after the virus is no longer detectable.

Smallpox (variola): Although no longer a threat, the variola virus was responsible for this highly communicable disease that is transmitted via the respiratory route through the bloodstream to internal organs. Replication in the epidermis eventually produces pus-filled vesicles and pitted lesions (**pox**) that are less prominent on the trunk than the extremities and face (**Figure 2A**).

Monkeypox: A rare viral disease caused by the monkeypox virus. The disease is similar to but less severe than smallpox and results in lymph node swelling.

Molluscum contagiosum: A mildly contagious but uncommon disease spread by direct or indirect contact with the molluscum contagiosum virus. The virus, which can infect any part of the skin, produces small, wart-like tumors.

Cutaneous warts: Several strains of papillomaviruses cause warts by infecting epithelial cells. The viruses stimulate excessive epithelial cell division usually resulting in benign tumors (**common warts** or **plantar warts**).

Parasitic Skin Infections

Cutaneous leishmaniasis: This disease arises from the bite of a sandfly infected with the protozoan *Leishmania major*. A skin ulcer develops (**Figure 2B**) due to immune system attack on the pathogen; healing can take up to a year.

Treatment

Antibiotics are ineffective against viral infections. Diseases like chickenpox, measles, and rubella are overcome by the immune system. Vaccines also are available. The antiviral drug acyclovir is useful in reducing the acute pain and time course of shingles and in shortening the healing time for cold sores. Other base analogs are effective against HSV in minimizing viral shedding and limiting recurrences.

Warts sometimes spontaneously disappear. They can be removed by cryotherapy with liquid nitrogen or by "tissue burning" with caustic chemical agents. Vaccination has eradicated smallpox worldwide. An antimony compound is effective against cutaneous leishmaniasis.

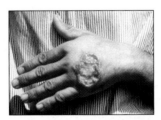

2B Cutaneous leishmaniasis. A skin ulcer on the hand of an adult with cutaneous leishmaniasis. (Courtesy of Dr. D.S. Martin/CDC)

2A Smallpox. A young boy who has contracted smallpox with the characteristic maculopapular rash on his face (not shown), torso, and arm. (Courtesy of Jean Roy/CDC)

3. Eye Infections

Background

Being exposed to the environment, the eyes represent a portal of entry and are subject to infection and disease (see right). Normally, eye infections are rare partly because the normal tear film, containing IgA antibodies, lysozyme, and other antimicrobial chemicals, continually flushes the eyes. Scratches or very dry eyes can decrease the effectiveness of these protective mechanisms and make infection more likely.

A few bacteria, viruses, and parasites are associated with eye infections. These agents can cause an inflammation of the **conjunctiva** (**conjunctivitis**), the thin mucous membrane that lines the inner surface of the eyelids and the white portion of the eyeball. They may also cause an inflammation of the **cornea** (**keratitis**), the transparent part of the eyeball covering the iris and the lens.

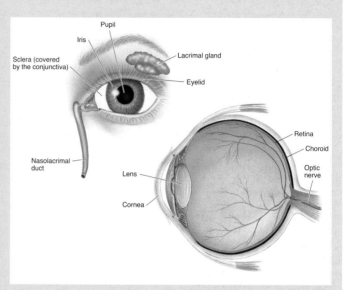

Signs and Symptoms

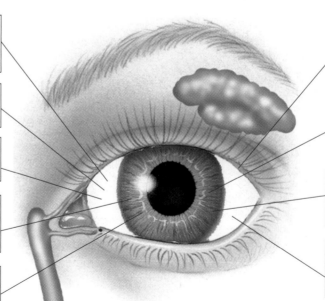

Bacterial and acute viral conjunctivitis: Reddening, itching, burning sensation, watery to sticky discharge.

Neonatal conjunctivitis: Pus-containing discharge, swelling of the conjunctiva and eyelid.

Adult inclusion conjunctivitis: Red, irritable eye with pus-containing discharge.

Trachoma: Tearing, mucous discharge, redness, sensitivity to bright light, corneal scarring, blindness.

Onchocerciasis: Chronic dermatitis, with ocular lesions often leading to blindness

Blepharitis: Redness, itching and burning sensations, swelling, and small abscess and scaling of eyelid margin.

***Acanthamoeba* keratitis:** Itching, tearing, watery discharge, decreased vision, sensitivity to light.

Infectious keratitis: Red eye with severe pain, tearing, sensitivity to light, and blurred vision.

Epidemic conjunctivitis: Red and swollen conjunctiva, with watery discharge; painful cornea.

Pathogenesis

The following is a brief description of human eye diseases caused by the most common pathogens.

Bacterial Diseases

Blepharitis: An acute or chronic inflammation, often caused by *Staphylococcus aureus*, at the base of an eyelash or blocking a sebaceous gland on the eyelid (**sty**).

Conjunctivitis

- **Bacterial:** Primarily caused by *Haemophilus influenzae*, this form of conjunctivitis is commonly called **pink eye (Figure 3A)**. Usually, there is no affect on vision.

- **Neonatal: Gonorrheal ophthalmia**, caused by *Neisseria gonorrhoeae*, and **chlamydial ophthalmia**, caused by *Chlamydia trachomatis*, are forms of an acute inflammation that can be passed from infected mother during birth. The inflammation can cause blood vessel dilation, conjunctiva edema, and excessive secretion **(Figure 3B)**.

- **Adult inclusion:** Caused by different serovars of *C. trachomatis*, this is mild in form and rarely causes blindness. It is spread from genitalia to the eye.

- **Trachoma:** A chronic form of keratoconjunctivitis caused by specific serotypes of *C. trachomatis*. The bacterial cells are spread to the eyes by contaminated hands, fomites, or flies. Infection causes an inflammation of the conjunctiva with eventual corneal scarring. Blindness can result. Trachoma is primarily found in developing nations; in fact, it is the greatest single infectious cause of adult blindness worldwide.

Viral Diseases
Conjunctivitis

- **Acute viral:** An infection usually caused by adenoviruses; also called **pink eye**, due to the resulting pinkness of the conjunctiva. The virus can be transmitted by hands, contaminated eye drops, or improperly chlorinated swimming pools.

- **Epidemic (keratoconjunctivitis):** A serious form of conjunctivitis caused by specific adenovirus serotypes often coming from improperly cleaned ophthalmologic instruments. It can result in corneal opacity for several years.

- **Infectious (herpetic) keratitis:** An infection of the cornea caused by herpes simplex virus-1 and -2. Inflammation starts in the conjunctiva but eventually affects the cornea, causing corneal ulcers.

Parasitic Diseases

***Acanthamoeba* keratitis:** A local infection of the eye caused by the protozoan *Acanthamoeba*. The disease is most prevalent in individuals wearing contact lenses that are not properly disinfected. Outbreaks have also been associated with improperly chlorinated swimming pools, communal sharing of infected towels, or ophthalmic procedures using contaminated equipment.

Onchocerciasis: A filarial infection of the skin and eye in tropical Africa, Mexico, and Central and South America, caused by *Onchocerca volvulus*. Carried by black flies, the larvae (microfilariae) can mature into adults that exist as fibrous nodules in the subcutaneous tissue. Larvae produced from the adults can migrate to the eyeball. Infection of the vitreous humor and ciliary body leads to visual impairment and eventually blindness (commonly called **"river blindness"**).

✚ Treatment

Blepharitis can be treated with warm compresses and perhaps a topical antibiotic. Both forms of neonatal conjunctivitis can be treated with erythromycin eye drops to prevent the onset of disease, while adult inclusion conjunctivitis involves treatment with a broad spectrum antibiotic. Trachoma is sensitive to several antibiotics, including erythromycin and tetracycline. However, these drugs only control cell growth, as they cannot eliminate the bacterial cells from the eye. Repeated infections lead to permanent scarring, and, if available, a corneal transplant is necessary to restore vision.

Acute viral conjunctivitis usually resolves with treatment; there are no antiviral drugs available for epidemic conjunctivitis. Infectious keratitis usually resolves on its own, although infections can be effectively treated with trifluridine eye drops. Several prescription eye medications are available for *Acanthamoeba* keratitis while ivermectin works well against *O. volvulus*.

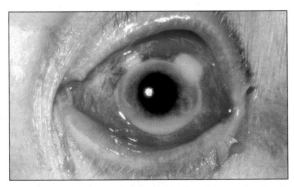

3A Bacterial Conjunctivitis. The reddening of the conjunctiva is obvious in this patient. Such an infection is often called pink eye. (© Medical-on-Line/Alamy)

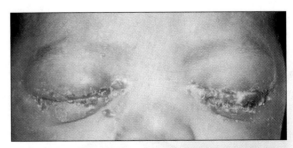

3B Neonatal Conjunctivitis. A newborn with gonorrhea ophthalmia resulting from contact with *Neisseria gonorrhoeae* during passage through the infected mother's birth canal. A purulent discharge typically forms in the eyes (Courtesy of J. Pledger/CDC)

4. Bacterial and Viral Infections

Background

The **upper respiratory tract** (**URT**) consists of the nose, sinus cavities, pharynx (throat), and larynx (see right). Since many the middle ear infections are the result of URT infections, they will be considered here as well. The normal microbiota of the URT suppresses colonization and inhibits pathogen growth while mucous membranes lining the nose and upper throat trap microbes. The **lower respiratory tract** (**LRT**) consists of the trachea, bronchi, and lungs. The LRT lacks much of the microbiota typical of the URT; therefore, infections can be more serious.

With regard to the URT, infections often are transmitted by airborne droplets or droplet nuclei from infected individuals or carriers. About 90% of URT infections are viral—most the result of cold viruses. URT infections exhibit typical inflammatory **syndromes** including sinusitis, rhinitis, epiglottitis, and pharyngitis (**Figure 4A**).

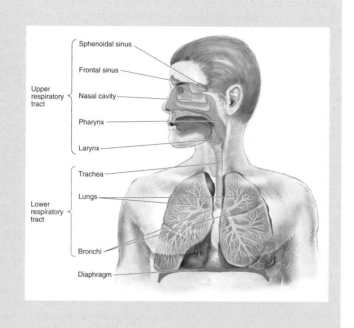

Signs and Symptoms

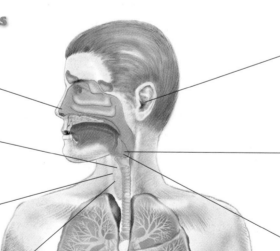

Common cold: Runny nose, headache, congestion, sneezing, and sore throat.

Epiglottitis: Severe throat pain, fever, muffled voice; stridor.

Strep throat: Sore throat, red swollen tonsils and pharynx, fever, swollen lymph nodes.

can lead to

Scarlet fever: High fever, swollen lymph nodes, sore throat; "strawberry" tongue followed by a diffuse red rash on neck, torso, and extremities.

Otitis media: Deep ear pain, impaired hearing, and bulging eardrum; fever, dizziness, and nausea.

Diphtheria: Moderate fever and sore throat; lesions on pharynx, larynx, and tonsils; grayish pseudomembrane on throat and nasopharynx.

Parainfluenza: Fever, profuse nasal discharge, sore throat, and dry, barking cough.

Other URT Infections

The bacterial species responsible for **meningococcal meningitis** (*Neisseria meningitidis*) and **Haemophilus** meningitis (*Haemophilus influenzae*) can enter by way of the URT. Both will be described with the diseases of the nervous system.

Pathogenesis

The following is a brief description of several bacterial and viral infections of the URT.

Bacterial Skin Diseases

Strep throat (streptococcal pharyngitis): This classic streptococcal infection caused by *Streptococcus pyogenes* (group A streptococci; GAS) occurs when the bacterial cells are transmitted person-to-person via droplets or nasal secretions (**Figure 4B**).

Scarlet fever: An initially local tonsillar or pharyngeal infection, the disease can spread as a body rash to the neck, trunk, and extremities after the release of pyrogenic exotoxins.

Epiglottitis: This is a serious infection (especially in children) often caused by *Haemophilus influenzae*. Inhaled in aerosols, colonization can lead to URT inflammation and swelling of the epiglottis and surrounding structures, which can completely block the airway through the trachea and into the lungs.

Diphtheria: A highly contagious and dangerous local infection of the pharyngeal mucous membranes (pharyngitis) caused by *Corynebacterium diphtheriae*. Infection from respiratory droplets produces an exotoxin that damages the throat, causing throat and lymph node swelling. Throat swelling can block the airway leading to suffocation. The toxin also affects the heart and peripheral nerves.

Otitis media: An acute bacterial infection of the middle ear caused by *Streptococcus pneumoniae* or *Haemophilus influenzae*. Contact with respiratory droplets and URT colonization, or reflux through the auditory tubes, leads to a middle ear infection most common in young children. Chronic otitis media (COM) can occur and cause damage to the middle ear.

Viral Diseases

Common cold (coryza): There are more than 200 virus subtypes that can cause a common cold, which is an infection of the nose, sinuses, and pharynx. Sinusitis and otitis media may be complications. Signs and symptoms can vary somewhat depending on the infecting virus.

- **Rhinoviruses:** More than 100 subtypes make the rhinoviruses (Picornaviridae), the most common cause of typical **"head colds."** The viruses infect the mucosal layer of the URT and attach to the nasal epithelium. Infection results in the release of more viruses that spread the infection in nasal discharges.

- **Coronaviruses.** Some members of this viral group (Coronaviridae) produce colds that are indistinguishable from the rhinoviruses.

- **Adenoviruses:** These viruses (Adenoviridae) also cause URT infections involving a substantial fever, an intense sore throat, and cough. Infection may spread to the lower respiratory tract.

- **Parainfluenza:** Inhaled through aerosols, subtypes 1, 2, and 3 cause URT infections in adults, often with typical cold symptoms. Human parainfluenza virus subtype 1 (HPIV-1) is the leading cause of **croup** (infection of the larynx and other URT structures) in children.

Treatment

Although strep throat itself is not usually dangerous, complications of rheumatic fever and kidney damage can be, so treatment with penicillin or erythromycin is necessary. For epiglottitis, once the patient is breathing smoothly, intravenous antibiotics can be used.

Prevention of diphtheria is accomplished through infant immunization with the DPT vaccine. Treatment involves administration of diphtheria antitoxin and antibiotic (penicillin or erythromycin) therapy. Antibiotic ear drops and oral antibiotics can be used to treat COM.

Because antibiotics are useless against viruses, common colds must just run their course. Over-the-counter (nonprescription) medications, including antihistamines and decongestants, may relieve symptoms.

4A URT syndromes. A variety of syndromes (signs and symptoms) can be associated with the URT.

Sinusitis
Rhinitis
Otitis media (middle ear)
Tonsillitis
Pharyngitis
Epiglottitis

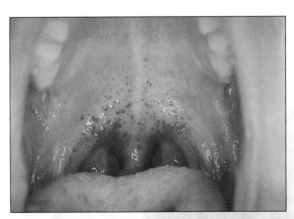

4B Strep throat. The infection is characterized by an inflammation of the oropharynx. Small red spots appear on the soft palate. (Courtesy of Dr. Heinz F. Eichenwald/CDC)

5. Bacterial Infections

Background

The lower respiratory tract (LRT) consists of the trachea (windpipe), bronchial tubes, and the lungs (see Unit 4). Due to the mucous membranes and filtering mechanisms of the bronchial tubes, the LRT normally contains a less dense microbiota population. Therefore, if endogenous or exogenous pathogens enter the LRT, serious respiratory disease may result.

Besides being a common **portal of entry**, the respiratory system is a **portal of exit**. Coughing and sneezing are two common ways pathogens can be transmitted via aerosols from person to person.

Infections of the LRT exhibit themselves in several ways based on their anatomical location (**Figure 5A**):

Bronchitis. A common infection of the bronchial lining, the inflammation produces a thick mucus that narrows the airways. Individuals, already with lung disease (often as a result of smoking), may develop chronic bronchitis, often referred to as **chronic obstructive pulmonary disease** (**COPD**).

Bronchiolitis. Usually restricted to young children, an infection of the bronchiole lining causes a swelling and narrowing of the airways, making expiration difficult (a wheezing sound is heard).

Pneumonia. An acute and complex syndrome resulting from an infection of the lung tissue and alveoli. Impaired gas exchange causes rapid and labored breathing and cough. Pneumonia can be community acquired (infection acquired outside a hospital) or hospital acquired (**nosocomial**). The syndrome is often divided into "**typical**" **pneumonia**, caused by the most common bacterial pathogens of the LRT (e.g., *Streptococcus pneumoniae*), and "**atypical**" **pneumonia,** which involves less common pathogens that usually cause a more mild disease.

Signs and Symptoms

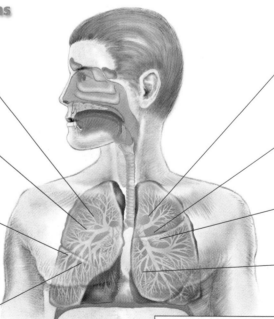

Inhalation anthrax: Fever, chills, cough, chest pain, headache, malaise; severe breathing and shock result.

Tuberculosis: Fever, fatigue, weight loss, cough; shortness of breath and chest pain; tubercle development

Primary atypical and chlamydial pneumonia: Gradual and mild symptoms with fever, fatigue, and dry, hacking cough.

Pertussis: *Catarrhal stage*: malaise, dry cough, fever; *Paroxysmal stage*: violent (whooping) cough; *Convalescent stage*: sporadic cough that slowly subsides.

Pneumococcal pneumonia: High fever, chest pain, persistent cough, rust-colored sputum; increased pulse and difficulty breathing.

Q Fever: Dry cough, high fever, chest pain, severe headache.

Legionellosis: Pneumonia symptoms with fever, dry cough, diarrhea, and vomiting.

Ornithosis: Fever, headache, and dry cough

Nosocomial Pneumonias
Species of *Citrobacter*, *Enterobacter*, *Pseudomonas*, and *Staphylococcus* may produce pneumonias through a hospital-acquired infection.

Pathogenesis

Bacterial diseases of the LRT are outlined below.

"Typical" Pneumonia

Pneumococcal pneumonia: This community-acquired pneumonia, caused by *Streptococcus pneumoniae*, starts after a viral infection of the URT damages the airways. The infection can travel to the alveoli, causing inflammation and **lobar pneumonia**. Without appropriate antibiotic treatment, mortality is high, especially in individuals over 65 years.

"Atypical" Pneumonias

"Atypical" (walking) pneumonia: Caused by *Mycoplasma pneumoniae*, the cells colonize the respiratory epithelium and lead to inflammation. The community-acquired disease is common in children and teenagers but is rarely fatal.

Chlamydial pneumonia: Caused by *Chlamydophila* (formerly *Chlamydia*) *pneumoniae*, the community-acquired pneumonia develops symptoms and outcomes similar to those caused by *Mycoplasma pneumoniae*.

Legionellosis (Legionnaires' disease): An atypical, community-acquired pneumonia caused by *Legionella pneumophila*. Inhaled aerosols from contaminated water systems lead to ingestion by alveolar macrophages in which the bacterial cells survive and reproduce. Microabscesses result leading to severe pneumonia, especially in older individuals.

Q fever: Another atypical, community-acquired pneumonia, caused by *Coxiella burnetii*, is transmitted by inhaling aerosol droplets or consuming contaminated meat or unpasteurized milk from infected animals. The disease produces a mild pneumonia with a low mortality rate.

Ornithosis (psittacosis): A rare, community-acquired pneumonia caused by the bacterium *Chlamydophila* (formerly *Chlamydia*) *psittaci*. The obligate intracellular bacterial cells are inhaled in dried droppings from infected birds (parrots, parakeets, pigeons, turkeys). In the alveoli, the cells are ingested by macrophages and cause alveolar hemorrhage.

Other Bacterial Diseases

Tuberculosis (TB): An infection by *Mycobacterium tuberculosis*, the infection starts by inhalaton of bacilli in aerosols from an infected person. In the alveoli, the bacilli are ingested by (but survive in) macrophages. Immune cells wall off the infection, leading to granuloma formation and leaving calcified scars (**tubercles**)(**Figure 5B**).

Pertussis (whooping cough): Caused by *Bordetella pertussis*, this highly contagious disease is transmitted in aerosols and produces exotoxins that inhibit normal respiratory clearance. A violent cough develops and straining for air causes a "whooping" sound.

Inhalation anthrax: This deadly disease is caused by the inhalation and germination of *Bacillus anthracis* spores. In the bloodstream, the bacterial cells divide and secrete exotoxins. If not treated early, inhalation anthrax is 100% fatal.

☤ Treatment

Penicillin is the drug of choice for pneumococcal pneumonia. Pneumococcal vaccines are available as a preventative measure. Atypical pneumonias can be treated through antibiotic therapy: "atypical" and chlamydial pneumonia with erythromycin or tetracycline; legionellosis with erythromycin; Q fever with doxycycline; and ornithosis with tetracycline.

TB treatment historically involved extended use of isoniazid and rifampin for 6 to 9 months. However, **multidrug-resistant TB (MDR-TB)** and **extensively drug-resistant TB (XDR-TB)** require second- and third-line antibiotics. Pertussis can be treated with erythromycin, and, if caught very early, inhalation anthrax is treatable with ciprofloxacin.

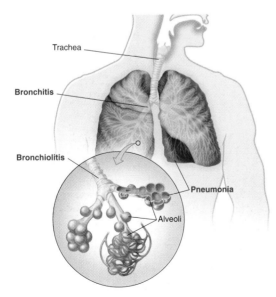

5A LRT syndromes. A variety of syndromes (signs and symptoms) can be associated with the LRT.

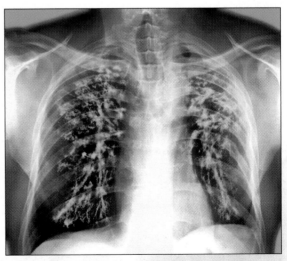

5B Tuberculosis. Colored chest X-ray showing lung scaring (green) caused by a chronic tuberculosis infection. (© Du Cane Medical Imaging Ltd./Photo Researchers, Inc.)

6. Viral and Fungal Infections

Viral infections of the lower respiratory tract (LRT) include influenza, which kills thousands of people worldwide each year, hantavirus pulmonary syndrome (HPS), and more recently SARS. Viruses also cause **viral pneumonias**, the most common such virus being the respiratory syncytial virus (RSV). These usually are the result of spread through the community, person to person.

Many of the serious fungal diseases of the LRT are **systemic mycoses**; that is, fungal diseases that can spread from the portal of entry to many body tissues. Organisms such as *Histoplasma capsulatum*, *Blastomyces dermatitidis*, and *Coccidioides immitis* primarily infect the lungs. Although most infections are asymptomatic and individuals recover without antimicrobial therapy, sometimes these fungi spread to secondary sites, causing disease that is more serious. Each type of mycosis is restricted to a specific geographical region. Also, these three fungi are

dimorphic, going through saprobic and parasitic phases that are temperature dependent.

The saprobic, filamentous phase of all three fungi grows in the soil where they form **conidiospores** or fragment into **arthrospores** (see below). Airborne transmission of these spores into the respiratory tract triggers the parasitic phase, which can lead to the formation of **yeast forms** or the development of **spherules**.

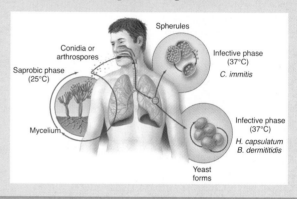

Signs and Symptoms

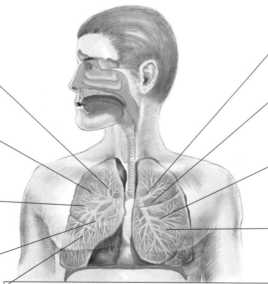

Coccidioidomycosis: *Acute form*: chest pain, fever, chills, dry cough, headache, and rash; *Chronic form*: fever, weight loss, chest pain, and blood in sputum.

Blastomycosis: Resembles a flu-like illness; the persistent form includes cough, chest pain, fever, chills, night sweats, and malaise.

Histoplasmosis: *Acute form*: fever, chest pains, headache, and cough; *Progressive form*: fatigue, malaise, and other symptoms mimicking TB.

Pneumocystis pneumonia: Shortness of breath, fever, dry cough, fatigue, weight loss.

Aspergillosis: Fever, chest pain, headache, shortness of breath, and bloody cough.

Influenza: Fever, chills, muscle aches, headache, dry cough, chest pains, and loss of appetite.

RS disease: *Infants*: high fever, wheezing, severe cough, and respiratory distress; *Older children and adults*: runny nose, cough, sore throat, malaise, and irritability.

SARS: Fever, headache, chills, malaise; developing into dry cough, pneumonia, and respiratory distress.

Hantavirus pulmonary syndrome: Fever and muscle aches; developing into cough and respiratory distress.

Other Fungal Diseases
Cryptococcosis is a potentially dangerous disease caused by *Cryptococcus neoformans* and *C. gattii*. Besides a pulmonary infection, the fungal species can cause disseminated infections, including meningitis. Symptoms include piercing headaches, neck stiffness, and paralysis, usually in immunocompromised individuals.

Pathogenesis

Viral and fungal diseases of the LRT are summarized below.

Viral Diseases

Influenza (flu): A viral infection caused by the influenzaviruses (Orthomyxoviridae). Transmission is by direct or indirect contact. The viruses infect the bronchi and bronchioles. Illness is often severe but only lasts for 3 to 5 days. Lung damage can lead to pneumonia.

Respiratory syncytial (RS) disease: RS disease is caused by the respiratory syncytial virus (RSV; Paramyxoviridae) and represents the most common cause of bronchitis, bronchiolitis, and **viral pneumonia** among infants and young children. The virus is transmitted by respiratory secretions.

Severe acute respiratory syndrome (SARS): This infectious disease is caused by SARS-CoV (Coronaviridae), which spreads via droplets (face-to-face contact) or airborne particles. In 10% to 20% of affected individuals, severe injury occurs in the lungs, requiring mechanical ventilation.

Hantavirus pulmonary syndrome (HPS): Caused by strains of hantavirus (Bunyaviridae), the infection usually occurs after contact with infected rodents or aerosolized rodent urine. Recent outbreaks in the southwestern United States lead to lung failure as capillaries become leaky and fluid fills the air spaces (**Figure 6A**).

Fungal Diseases

Coccidioidomycosis (valley fever): Caused by *C. immitis* and *C. posadasii*, the airborne conidia (**arthrospores**) are inhaled. In the bronchioles and alveoli the spores form thick-walled **spherules**. Usually, the acute primary disease is without symptoms. The rare, chronic and progressive disease, usually in people with a weakened immune system, produces abscesses throughout the body, which can be life threatening.

Histoplasmosis: The fungus *H. capsulatum* is the etiologic agent for this most common fungal respiratory disease. Aerosolized bird or bat feces contain fungal spores that are inhaled. The acute disease can be asymptomatic or symptomatic pneumonia. Without recovery, the chronic form leads to progressive pulmonary disease.

Blastomycosis: Inhalation of *B. dermatitidis* spores leads to an infection of the lungs and produces bronchopneumonia. The disease can spread to other parts of the body, including the skin where papules or pustules form.

***Pneumocystis* pneumonia (PCP):** This opportunistic infection by *Pneumocystis jirovecii* (formerly *Pneumocystis carinii*) comes from the fungus already in the lungs as part of the human microbiota. Lung invasion leads to microbe and exudate-filled alveoli, resulting in a loss of gas exchange. PCP occurs in some 90% of HIV-infected patients.

Aspergillosis: An opportunistic infection resulting from inhalation of *Aspergillus fumigatus* conidia from the environment. In individuals suffering a prior disease (e.g., tuberculosis), germination of the conidia in the lungs can produce a noninvasive "fungus ball" (**aspergilloma**; **Figure 6B**). **Invasive aspergillosis** can spread from the lung to the brain and other body organs.

⚕ Treatment

Antibiotics are of no use against viral infections. Bed rest, plenty of fluids, and over-the-counter medications usually are sufficient for the flu. In some cases, the antivirals oseltamivir (Tamiflu) or zanamivir (Relenza) may be prescribed to lessen the symptoms. For severe RS disease, hospitalization for intravenous (IV) fluids and humidified oxygen may be needed. No effective antivirals exist for SARS, although a combination of lopinavir, ritonavir, and ribavirin may prevent serious complications and death. HPS treatment involves supportive therapy and, for severe cases, blood oxygenation.

The antifungal drugs amphotericin B and itraconazole are used for the more serious forms of valley fever, histoplasmosis, and blastomycosis. Treatment of PCP involves trimethoprim-sulfamethoxazole, which has very toxic side effects. Voriconazole can be used to treat invasive aspergillosis.

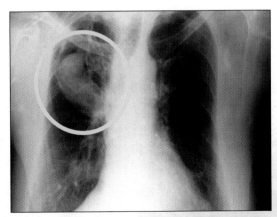

6A Hantavirus pulmonary syndrome. A chest X-ray image showing the pulmonary effusion (oval) caused by HPS. (Courtesy of D. Loren Ketai, M.D./CDC)

6B Aspergillosis. This X-ray image shows the development of an aspergilloma (circle) in the upper lobe of the right lung. (Courtesy of M. Renz/CDC)

7. Bacterial Infections

Background

The **digestive system** consists of the **gastrointestinal (GI) tract** and the **accessory digestive organs** (see right). Because much of the digestive system is a tube running through the body from mouth to anus, these areas are colonized by many normal microbiota species.

Infections can be separated by the clinical **syndromes** in the oral cavity or digestive tract (**Figure 7a**):

Gingivitis. An inflammation of the gums often resulting from inadequate tooth brushing and flossing.

Periodontitis. A severe form of gingivitis extending into the supporting structures of the tooth.

Gastritis. An inflammation of the stomach lining.

Gastroenteritis. An often uncomfortable inflammation of the stomach or intestines, causing vomiting and diarrhea.

Colitis. An inflammation of the colon (large intestine), causing lower bowel spasms and abdominal cramps.

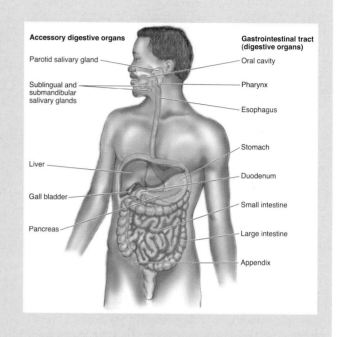

Accessory digestive organs
- Parotid salivary gland
- Sublingual and submandibular salivary glands
- Liver
- Gall bladder
- Pancreas

Gastrointestinal tract (digestive organs)
- Oral cavity
- Pharynx
- Esophagus
- Stomach
- Duodenum
- Small intestine
- Large intestine
- Appendix

Signs and Symptoms

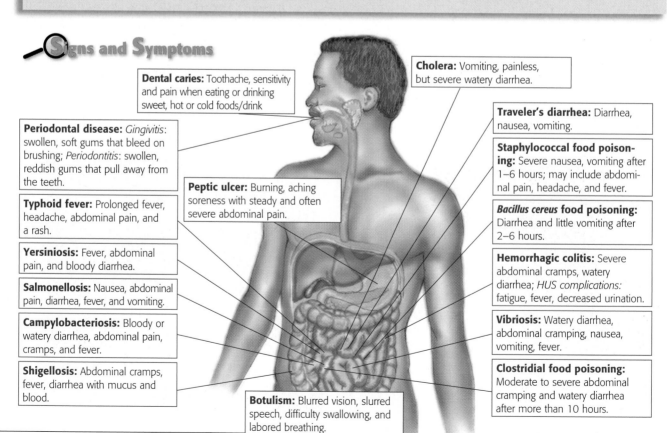

Dental caries: Toothache, sensitivity and pain when eating or drinking sweet, hot or cold foods/drink

Periodontal disease: *Gingivitis*: swollen, soft gums that bleed on brushing; *Periodontitis*: swollen, reddish gums that pull away from the teeth.

Typhoid fever: Prolonged fever, headache, abdominal pain, and a rash.

Yersiniosis: Fever, abdominal pain, and bloody diarrhea.

Salmonellosis: Nausea, abdominal pain, diarrhea, fever, and vomiting.

Campylobacteriosis: Bloody or watery diarrhea, abdominal pain, cramps, and fever.

Shigellosis: Abdominal cramps, fever, diarrhea with mucus and blood.

Peptic ulcer: Burning, aching soreness with steady and often severe abdominal pain.

Botulism: Blurred vision, slurred speech, difficulty swallowing, and labored breathing.

Cholera: Vomiting, painless, but severe watery diarrhea.

Traveler's diarrhea: Diarrhea, nausea, vomiting.

Staphylococcal food poisoning: Severe nausea, vomiting after 1–6 hours; may include abdominal pain, headache, and fever.

***Bacillus cereus* food poisoning:** Diarrhea and little vomiting after 2–6 hours.

Hemorrhagic colitis: Severe abdominal cramps, watery diarrhea; *HUS complications*: fatigue, fever, decreased urination.

Vibriosis: Watery diarrhea, abdominal cramping, nausea, vomiting, fever.

Clostridial food poisoning: Moderate to severe abdominal cramping and watery diarrhea after more than 10 hours.

Pathogenesis

Bacterial diseases of the digestive system are outlined below.

Oral Diseases

Dental caries: Tooth decay results from colonization of *Streptococcus mutans* and *S. sanguis* on the tooth pellicle. Fermentation with acid production can lead to a demineralization of the tooth enamel and, progressing into the pulp, a painful cavity (caries).

Periodontal disease: This gum inflammation includes:

- **Gingivitis.** An early-stage periodontal disease, often caused by species of *Prevotella* and *Fusobacterium*, is common among young adults not brushing their teeth.

- **Periodontitis.** A serious disease resulting from untreated gingivitis. The infection can lead to bone resorption and tooth loss.

Gastritis

Peptic ulcer: An open sore on the inside lining of the stomach (gastric ulcer) or duodenum (duodenal ulcer) caused by *Helicobacter pylori*. Inflammation leads to ulcer formation, which, if left untreated, can cause internal bleeding and infection (**peritonitis**).

Noninflammatory Gastroenteritis

Staphylococcal food poisoning: The intoxication comes from ingesting *S. aureus* **enterotoxins** in contaminated food. The toxins trigger substantial diarrhea.

Clostridial food poisoning: An intoxication caused by the germination of *Clostridium perfringens* spores transmitted in contaminated meat and poultry. A secreted enterotoxin damages the epithelium, producing abdominal cramps and diarrhea.

***Bacillus cereus* food poisoning:** Another intoxication, caused by *Bacillus cereus*, results from ingestion of contaminated food. Two enterotoxins cause vomiting and diarrhea.

Botulism: A very dangerous intoxication caused by *Clostridium botulinum*. Spores in improperly canned foods can germinate and the vegetative cells produce a **neurotoxin** that when ingested quickly can lead to respiratory paralysis and death.

Inflammatory Gastroenteritis

Cholera: A diarrheal disease caused by *Vibrio cholerae*. Drinking contaminated water transmits the pathogen to the small intestine, where it produces an enterotoxin that stimulates an unrelenting loss of water and electrolytes. Severe dehydration and death can occur without treatment.

Traveler's diarrhea: An infection caused by **enterotoxigenic** *Escherichia coli* (**ETEC**). The noninvasive cells produce enterotoxins in the small intestine that stimulate a watery diarrhea.

Vibriosos: A food illness caused by *Vibrio parahaemolyticus* and *Vibrio vulnificus*, halophiles associated with contaminated shellfish. Ingestion by immunocompromised individuals can lead to a systemic infection, and the mortality rate can be as high as 50%.

Invasive Gastroenteritis

Typhoid (enteric) fever: A disease caused by *Salmonella enterica* serotype Typhi. Fecal-oral transmission leads to infection of the gall bladder and liver and endotoxin release. Blood invasion causes a fever and development of a chest and abdominal rash (**rose spots**). After recovery, persons may remain carriers.

Salmonellosis: An infection most commonly caused by *S. enterica* serotype Enteritidis and *S. enterica* serotype Typhimurium contaminating milk and poultry products.

Shigellosis: An infection from food contaminated with *Shigella sonnei*. The enterotoxin triggers gastroenteritis in the distal ileum or colon.

Hemorrhagic colitis: A potentially severe disease caused by cytotoxins from **enterohemorrhagic** *E. coli* (**EHEC**) O157:H7 that damage the lining of the colon. Dissemination to the kidneys leads to **hemorrhagic uremic syndrome** (**HUS**).

Campylobacteriosis: An infection caused by transmission of *Campylobacter jejuni* contaminating food or unpasteurized milk. The cells invade and damage the mucosal surfaces of the small intestine and colon.

Yersiniosis: An emerging foodborne illness caused by *Yersinia enterocolitica*. The species inhabits domestic animals, and consumption of contaminated food can, in children, cause endotoxin-induced tissue destruction in the ileum.

✚ Treatment

For most cases of bacterial gastroenteritis, antibiotics either are not given or are given to shorten the duration of the illness and decrease the likelihood of transmission. In individuals with food poisoning or disease with diarrhea, fluid consumption is important to prevent dehydration.

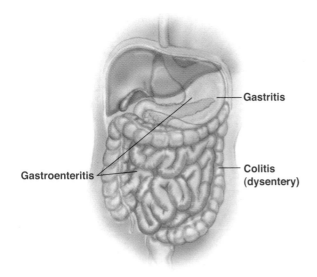

Gastritis

Gastroenteritis

Colitis (dysentery)

7A Digestive system syndromes. A few syndromes (signs and symptoms) associated with the GI tract.

8. Viral Infections and Fungal Intoxications

Background

Besides bacteria, viruses can affect the digestive system. In fact, viruses are more common than bacterial species in causing gastroenteritis.

Viral gastroenteritis can occur in many individuals. The best understood are infections with rotaviruses (infants and children) and noroviruses (children and adults). Rotavirus infections can cause severe dehydration and, as such, are a major cause of infant mortality in developing nations. Rotavirus infections also account for 50% of infant hospitalizations for diarrhea in the United States. The noroviruses usually cause self-limiting but explosive outbreaks (e.g., nursing homes, cruise ships) lasting 24 to 48 hours.

Several distinct viruses can cause **hepatitis**, an inflammation of the liver. They all have fairly similar clinical presentations. An **acute infection** usually has a sudden but short inflammatory period of a few weeks, while a **chronic infection** can persist for six months or longer.

Other viral diseases can involve the digestive system. For example, the enteroviruses (coxsackie A and coxsackie B viruses; Picornaviridae) are transmitted via the fecal-oral route and travel in the GI tract. However, they then spread to other tissues and can cause a variety of diseases, including aseptic meningitis, pericarditis, or hand-foot-and-mouth disease.

Although fungal organisms are not usually of concern as infectious whole agents responsible for digestive system disease, products of fungal metabolism can cause various illnesses. Some molds produce **mycotoxins**, the intoxication arising from the ingestion of the mycotoxin in food. Besides toxigenic molds, some mushroom species produce toxins that can have a variety of effects (**mycotoxicoses**) on human hosts, including digestive disorders from ingesting poisonous mushrooms (**mycetism**).

 **Signs and Symptoms**

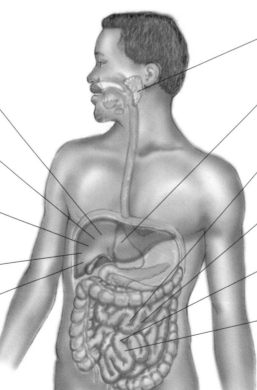

Hepatitis A: Often asymptomatic; fever, anorexia, nausea, vomiting, fatigue, and abdominal pain; followed by jaundice.

Hepatitis B: Anorexia, nausea, vomiting, fatigue, joint pain, and often a rash; progressing to jaundice.

Hepatitis D: Occurs as a co-infection with hepatitis B; signs and symptoms similar to hepatitis B.

Hepatitis C: Often asymptomatic; anorexia, nausea, vomiting, and malaise; progressing to jaundice (less frequent than in hepatitis B).

Hepatitis E: Similar to hepatitis A.

Mumps: Chills, headache, muscle aches, and fever; swelling of one or more salivary glands.

Aflatoxin poisoning: Vomiting, abdominal pain, convulsions, and fever.

Ergotism: A burning skin sensation, itching, headache, diarrhea, nausea, and vomiting; precede hallucinations and psychoses.

Mycetism: Vomiting, diarrhea, abdominal cramps.

Rotaviral gastroenteritis: Fever, watery diarrhea, abdominal pain, and vomiting.

Noroviral gastroenteritis: Fever, nausea, diarrhea, vomiting, stomach pain, malaise, and headache.

Pathogenesis

Viral diseases and fungal intoxications of the digestive system are outlined below.

Mumps: An acute, highly contagious viral disease of one or more salivary glands (most often the parotid gland; **parotitis**) caused by the mumps virus (Paramyxoviridae). Transmitted by saliva, the virus spreads via the URT to the lymph nodes and then via the blood to the salivary glands causing inflammation and edema. In about 25% of postpubertal males, the virus also spreads to the testes causing a painful enlargement (**orchitis**).

Viral Gastroenteritis

Rotaviral gastroenteritis: An often severe form of gastroenteritis in infants and young children caused by the human rotaviruses (Reoviridae). Transmitted by the fecal-oral route, the virus infects the proximal small intestines where it kills the host cells during replication. Cell death leads to impaired absorption of nutrients and results in vomiting and (often profuse) watery diarrhea.

Noroviral gastroenteritis: A mild-to-moderate disease caused by the highly contagious noroviruses (formerly called the Norwalk-like viruses; Caliciviridae). The viruses spread via the fecal-oral route. Although infection of the proximal small intestine leads to inflammation, knowledge of pathogenesis is unclear.

Hepatitis

Hepatitis A (infectious hepatitis): An often acute, but mild, liver infection caused by the hepatitis A virus (HAV; Picornaviridae). Spread via the fecal-oral route, HAV infects and replicates in mucosal epithelial cells of the GI tract. Blood spread brings the viruses to the liver where they infect hepatocytes. Affected individuals mount a cytotoxic T-cell response that usually brings complete recovery.

Hepatitis B (serum hepatitis): An acute infection caused by the hepatitis B virus (HBV; Hepadnaviridae). Transmitted through blood or sexual contact, HBV infects and replicates in hepatocytes. Cytotoxic T-cell activation leads to inflammation. In immunocompromised individuals, the infected hepatocytes are not cleared and **chronic hepatitis** occurs. In 15% to 25% of adult chronic cases (70% to 90% of infants), **cirrhosis** (destruction of liver tissue) or liver cancer (**hepatocellular carcinoma**) can develop (**Figure 8**).

Hepatitis C: An acute illness caused by the hepatitis C virus (HCV; Flaviviridae). HCV usually is transmitted via the sharing of contaminated needles and the viruses infect hepatocytes. Cytotoxic T cells kill infected cells. In about 75% of cases, the viruses are not eliminated and **chronic hepatitis** develops. Cirrhosis predisposes the individual to liver cancer.

Hepatitis D: An abrupt illness caused by a virus-like particle called hepatitis D virus (HDV). Transmitted through blood or sexual contact, HDV infects hepatocytes and replicates in those cells also infected with HBV. Hepatitis in such an individual is more severe than an independent hepatitis B infection.

Hepatitis E: An acute infection caused by the hepatitis E virus (HEV; Caliciviridae). Transmitted via the fecal-oral route, HEV infects mucosal epithelial cells where replication produces viruses that spread to the hepatocytes. Infected cells are eliminated by cytotoxic T cells.

Fungal Intoxications

Aflatoxin poisoning: An intoxication resulting from mycotoxins produced by the mold *Aspergillus flavus* that has contaminated grain before harvest or during storage. Ingested toxins are metabolized by the liver to a reactive product that can cause liver damage. They are potent **carcinogens** (cancer-causing agents) that can cause liver cell tumors.

Ergotism: An intoxication resulting from *Claviceps purpurea*. The fungus infects cereal grains (rye and wheat), during which time the alkaloid mycotoxin is produced. Ergotism arises from ingestion of these contaminated grain products. The toxin restricts blood flow at the limb extremities and causes hallucinations (psychotropic effects).

Mycetism (mushroom poisoning): An intoxication resulting from the ingestion of several species of mushrooms, especially the genus *Amanita*. These species produce very dangerous toxins (phalloidin and amanitin) that can cause liver damage. Without treatment, about 50% of individuals who ingest phalloidin die in 5 to 8 days.

Treatment

There is no treatment for mumps as there is the trivalent MMR vaccine.

For viral gastroenteritis, antibiotics and antiviral drugs are not prescribed; rehydration may be needed. For hepatitis: supportive care or immune globulin (HAV); interferon-alpha (HBV; HDV); interferon-alpha and ribavirin (HCV). No treatment has been developed for HEV.

Treatment of fungal intoxications involves attempts to decrease the amount of toxin in the body.

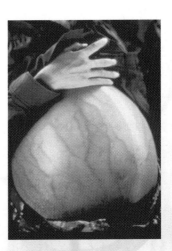

8 Hepatocellular carcinoma. This woman's distended abdomen is due to liver cancer resulting from a chronic hepatitis B infection. (Courtesy of CDC. Used with permission of Patricia F. Walker, M.D., D.T.M. & H., HealthPartners Center for International Health & International Travel Clinics)

9. Parasitic Infections

Parasites are organisms that live inside or on another organism, the **host**, and cause harm to the host. These parasites may be unicellular—the protozoa—or multicellular—the helminths. Many of these parasites can enter the body through the mouth during the ingestion of food or water contaminated with the parasite. Depending on the species, they may remain and reproduce in the intestines or spread to other organs of the body.

Parasitic diseases are most common in the developing parts of the world, where food and water can often become contaminated with the parasites. This is why American travelers to such areas are encouraged by the Centers for Disease Control and Prevention (CDC) to "cook it, boil it, peel it, or forget it."

Besides bacteria and viruses, a few protozoal species also cause **gastroenteritis**. Although the infections may differ in the site of infection, severity, and consequences,

these eukaryotic microbes enter the body as resistant **cysts** that can survive harsh environmental conditions for some period of time. Once in an appropriate host, the cysts develop into **trophozoites**, the feeding stage that often is responsible for digestive system disease. After completing their life cycle in the human digestive tract, more cysts are shed into the environment.

The helminths are worms, some of which are parasitic in humans. This includes the **flatworms**, such as the flukes and tapeworms, and the **roundworms**, the latter completing their entire life cycle in the human host. The helminths may do little damage in the intestine and so produce few symptoms. However, the parasites many disseminate to another part of the body and cause severe damage to vital organs. More than 70% of the world's population is infected with a parasitic helminth. Even in the United States, 30% of children and 16% of adults have enterobiasis (pinworm disease).

Signs and Symptoms

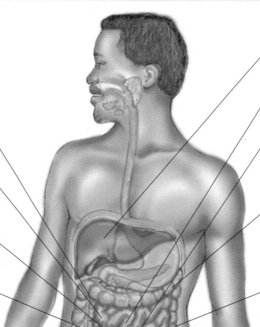

Amoebiasis: Abdominal pain and weight loss; bloody diarrhea in severe cases.

Giardiasis: Nausea, foul-smelling (fatty) diarrhea, abdominal cramping, and flatulence.

Cryptosporidiosis: Profuse, chronic, watery diarrhea, and abdominal cramps.

Cyclosporiasis: Watery diarrhea, nausea, abdominal pain, vomiting.

Echinococcosis: Asymptomatic.

Taeniasis: Asymptomatic, or mild diarrhea and abdominal discomfort.

Hookworm Disease: Itching, rash, anemia, abdominal pain, diarrhea, and weight loss

Ascariasis: Asymptomatic or abdominal pain with heavy worm loads.

Trichuriasis: Abdominal pain and anemia.

Enterobiasis: Perianal itching.

Other Helminth Diseases of the Digestive System:
Additional human parasitic diseases come from ingesting raw or undercooked fish containing larvae of *Diphyllobothrium latum* (fish tapeworm) and through cutaneous contact with larvae of *Strongyloides stercolaris* (threadworm) that eventually infect the intestines and cause gastroenteritis.

Pathogenesis

Protozoal and helminthic diseases of the digestive system are described below.

Protozoal Diseases

Amoebiasis: An inflammation of the colon caused by *Entamoeba histolytica* cysts in fecally contaminated food or water. Ingested cysts enter the small intestine and hatch out as trophozoites that then pass to the colon. In the colon, they cause a local necrosis leading to amebic dysentery. If trophozoites invade through the colon (**invasive amoebiasis**), they enter the blood circulation and travel to the liver and form abscesses.

Giardiasis: An often mild infection of the upper small intestine caused by drinking water contaminated with *Giardia intestinalis* (*G. lamblia*) cysts. In the duodenum, the cysts form trophozoites that attach to and damage the mucosa (inflammation). No further invasion occurs.

Cryptosporidiosis: An infection of the lower small intestine caused by drinking water contaminated with *Cryptosporidium parvum* or *C. hominis* cysts. The cysts release sporozoites in the intestine that develop into trophozoites that attach to the microvilli. In immunocompromised individuals, infection can cause a severe diarrhea that can be life threatening.

Cyclosporiasis: An infection of the intestines caused by the ingestion of fresh produce or water contaminated with oocysts of *Cyclospora cayetanensis*. In the GI tract, oocysts develop into sporozoites that invade the epithelium of the small intestine. In these cells, the parasite undergoes asexual and then sexual reproduction, forming more cysts.

Helminthic Diseases

Taeniasis: The two forms of this disease are caused by *Taenia saginata* (beef tapeworm) and *T. solium* (pork tapeworm). The tapeworm larvae are transmitted to humans in contaminated raw or undercooked beef and pork, respectively. In the intestines, the larvae grow into adult tapeworms that consume food ingested by the host (**Figure 9A**). Thus, they can interfere with normal physiological processes in the small intestine.

Echinococcosis: This infection is caused by the human ingestion of *Echinococcus granulosus* eggs in dog feces or in that of another carnivorous animal. Tapeworm eggs then hatch in the small intestine, and the larvae produced penetrate the intestinal wall and travel to the liver, lungs, or brain where **hydatid cysts** form. Cyst growth leads to organ dysfunction. Rupture of cysts can cause anaphylactic shock.

Enterobiasis (pinworm disease): A disease caused by the ingestion of eggs from the roundworm *Enterobius vermicularis*. Eggs hatch in the small intestine and mature into adults in the colon. Night migration of the female worm to the anus results in the deposition of eggs in a gelatinous material.

Ascariasis: This helminthic infection results from the ingestion of soil containing *Ascaris lumbricoides* eggs. Larvae hatch in the small intestine, invade the bloodstream, and move to the lungs. They then ascend to the trachea, causing inflammation and pneumonia. Moving from the trachea to the pharynx, the larvae are swallowed and in the small intestine mature into adults, where the worms live on partially digested food.

Trichuriasis (whipworm disease): A roundworm infection resulting from ingestion of soil containing eggs of *Trichuris trichiura*. The eggs hatch in the small intestine and migrate to the colon where they embed in the intestinal lining.

Hookworm disease: A common, chronic infection caused by *Necator americanus* and *Ancylostoma duodenale*. Larvae enter through the skin of the foot (**Figure 9B**), and, after traveling through the blood and lungs, they are passed to the small intestine, where they feed on blood and tissue.

✚ Treatment

Amoebiasis and giardiasis can be treated with antiprotozoal drugs, such as metronidazole. No effective treatment exists for cryptosporidiosis. Trimethoprim-sulfamethoxazole is effective against cyclosporiasis.

The beef and pork tapeworms (taeniasis) can be treated with praziquantal, while echinococcosis involves cyst surgery and treatment with albendazole. Roundworm infections can be treated with mebendazole.

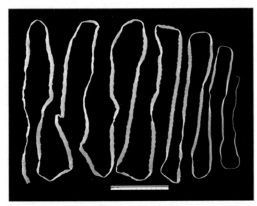

9A An adult tapeworm of *Taenia saginata*. Humans ingesting raw or undercooked infected meat can be infected with cysts that in about two months mature into adult tapeworms in the small intestine. (Courtesy of CDC)

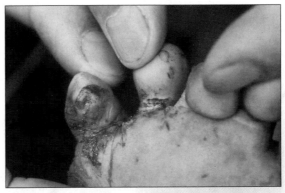

9B Hookworm infection. The toes of this patient's right foot are infected with hookworms. (Courtesy of CDC)

10. Bacterial, Fungal, and Protozoal Infections

Background

The nervous system is made up of the **central nervous system** (**CNS**), which consists of the brain and spinal cord, and the **peripheral nervous system**, which consists of the nerves extending from the CNS (see right). Although rare, infections of the CNS can be very serious, especially for infants and older adults. Infections may result in:

Meningitis. An inflammation of the membranes surrounding the brain and spinal cord (the **meninges**). Such infections by bacterial species or viruses cause swelling and increased pressure on the brain.

Encephalitis. An inflammation of the brain tissue that is usually caused by certain groups of viruses (see next unit). Such infections produce behavioral changes.

Myelitis. An inflammation of the spinal cord by bacterial species or viruses.

Infections of the CNS usually are the result of a complication from another infection, including pneumonia, sinusitis, or otitis media. The agents of these diseases reach the brain via the blood or from the sinus or middle ear infection. Infections also can be the result of a skull injury (fracture), head wound, or surgical procedure that exposes the CNS.

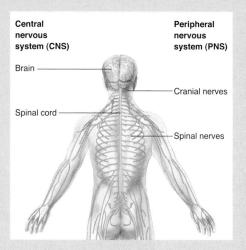

Signs and Symptoms

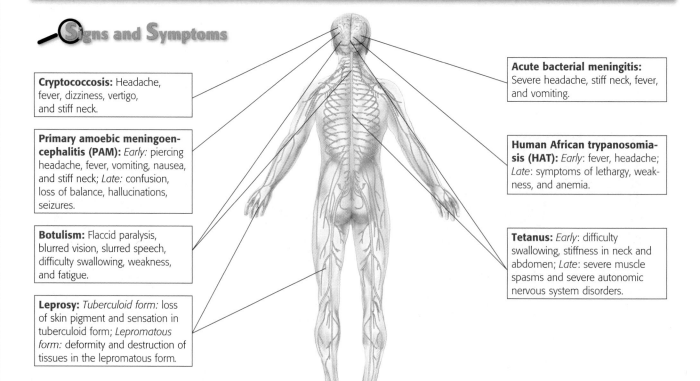

Cryptococcosis: Headache, fever, dizziness, vertigo, and stiff neck.

Primary amoebic meningoencephalitis (PAM): *Early:* piercing headache, fever, vomiting, nausea, and stiff neck; *Late:* confusion, loss of balance, hallucinations, seizures.

Botulism: Flaccid paralysis, blurred vision, slurred speech, difficulty swallowing, weakness, and fatigue.

Leprosy: *Tuberculoid form:* loss of skin pigment and sensation in tuberculoid form; *Lepromatous form:* deformity and destruction of tissues in the lepromatous form.

Acute bacterial meningitis: Severe headache, stiff neck, fever, and vomiting.

Human African trypanosomiasis (HAT): *Early:* fever, headache; *Late:* symptoms of lethargy, weakness, and anemia.

Tetanus: *Early:* difficulty swallowing, stiffness in neck and abdomen; *Late:* severe muscle spasms and severe autonomic nervous system disorders.

Pathogenesis

Bacterial, fungal, and protozoal diseases of the CNS are described below.

Acute Bacterial Meningitis

Meningococcal meningitis: An infection by *Neisseria meningitidis* that usually affects immunocompromised individuals. After colonizing the nasopharynx, the pathogen enters the bloodstream (**meningococcemia**). A released toxin causes a skin rash. Infection of the meninges causes inflammation and meningitis.

Haemophilus meningitis: An infection by *Haemophilus influenzae* (type B) in the bloodstream can spread to the CNS and lead to meningitis.

Pneumococcal meningitis: An infection caused by *Streptococcus pneumoniae*. Respiratory droplets first colonize the lung lymphatics before entering the bloodstream. Inflammation and meningitis can result from infection of the meninges (**Figure 10A**).

Listeric meningitis (listeriosis): A CNS disease caused by *Listeria monocytogenes* infection preferentially in immunocompromised individuals and pregnant women. The pathogen invades phagocytes of the GI mucosa. Spreading through the blood, it can target neural tissues and cause meningitis.

Other Forms of Meningitis

Cryptococcal meningitis (Cryptococcosis): An opportunistic infection by the yeast-like fungi *Cryptococcus neoformans* and *C. gattii* that affects persons with an impaired immune system. Inhalation of the organism leads to its spreading in the blood and to the CNS, causing meningitis. Damage results from abscess formation and tissue displacement.

Primary amoebic meningoencephalitis (PAM): An acute disease caused by the protozoan *Naegleria fowleri* or a more chronic illness due to *Acanthamoeba*. The amoeba usually enters the body through the nose from the individual swimming in warm freshwater. The pathogen then travels to the brain and spinal cord where it destroys brain tissue. Most patients die from the infection.

Other Bacterial Diseases

Botulism: This severe form of food poisoning in adults is caused the neurotoxins produced by *Clostridium botulinum* that has contaminated a food product. Nerve transmission is blocked because of toxin inhibition of acetylcholine release at nerve endings (**Figure 10B**).

Tetanus: This systemic disease is caused by a neurotoxin produced by *Clostridium tetani*. Spores enter through a skin injury, and the spores germinate. Vegetative cells release a neurotoxin that blocks inhibitory nerve transmitters. The result is sustained muscle contraction (also called **lockjaw**).

Leprosy (Hansen's disease): A chronic bacterial disease caused by *Mycobacterium leprae*. Bacterial multiplication in the peripheral nerves and skin macrophages leads to a tuberculoid (immunocompetent person) or lepromatous form (immunocompromised person); the latter produces inflammatory damage and sensory loss at extremities and face, resulting in disfiguring skin lesions (**lepromas**).

Protozoal Diseases

Human African trypanosomiasis (HAT): A systemic disease spread by the bite of a tsetse fly infected with either of two subspecies of *Trypanosoma brucei*. The pathogen multiplies in the blood, and, if it can escape the immune response, infection of the CNS causes encephalitis. One characteristic is lethargy, thus the common term **African sleeping sickness**.

⚕ Treatment

Bacterial meningitis requires quick action and therapy to prevent disabilities and death. Antibiotic treatment (ampicillin, cephalosporins) can be used on susceptible bacteria. Persons infected with *C. neoformans* and *C. gattii* can be treated effectively with amphotericin B. There is no treatment for PAM.

Treatment for botulism requires stomach pumping and antitoxin therapy. Tetanus requires immune globulin. Leprosy can be controlled with rifampin and dapsone. For the blood/lymph stage of HAT, suramin can be used; for the CNS stage, melarsoprol can be used.

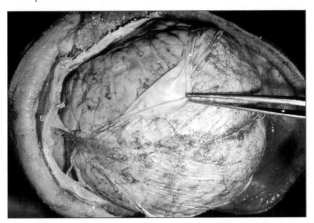

10A Pneumococcal meningitis at autopsy. A purulent inflammation is evident. (Courtesy of Dr. Edwin P. Ewing, Jr./CDC)

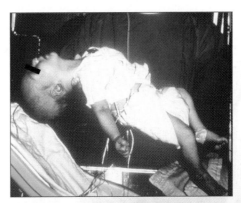

10B Botulism. This six-week old infant has botulism, as indicated by the loss of muscle tone. (Courtesy of CDC)

11. Viral and Viral-Like Infections

There are several viral diseases of the CNS. These viruses enter the CNS from the blood or peripheral axons. Most often, the viruses cause **aseptic meningitis**, a term used to describe meningitis when no bacterial cause can be found. Such cases of **viral meningitis** are usually mild and rarely life threatening.

Viruses also can cause **encephalitis** and **myelitis**. Such viral inflammations, especially encephalitis, can produce devastating and sometimes fatal results. For example, left untreated, some 70% of herpes encephalitis cases are fatal and almost all of the survivors suffer permanent neurological damage.

Some infections occur in epidemics, such as cases of aseptic meningitis caused by the echoviruses and Coxsackie viruses. In addition, the mumps and chickenpox viruses can cause isolated cases of encephalitis or myelitis.

Infection by the human immunodeficiency virus (HIV) can produce a chronic infection of the brain without characteristic acute encephalitis. Such situations are referred to as **AIDS dementia**.

Many cases of encephalitis are caused by the **arboviruses** (*ar*thropod-*bo*rne viruses), which are spread by mosquitoes, ticks, or other arthropods. The rabies virus is transmitted by the bite of a rabid animal, while lymphocytic choriomeningitis (LCM) is spread by the exposure to viruses in rodent feces or urine.

There also are diseases called **transmissible spongiform encephalopathies** (**TSEs**) caused by virus-like agents, specifically proteins called **prions**. The human form of the disease is called **variant Creutzfeldt-Jacob disease** (**vCJD**). Humans (primarily people living in the United Kingdom) have acquired these infectious agents through the consumption of prion-contaminated beef (**"mad cow" disease**).

Signs and Symptoms

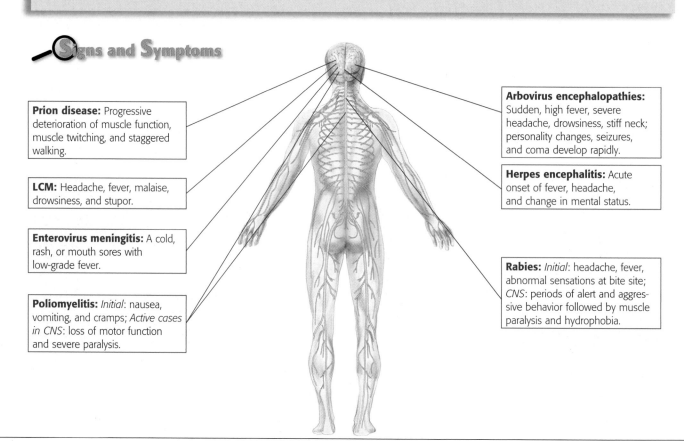

Prion disease: Progressive deterioration of muscle function, muscle twitching, and staggered walking.

LCM: Headache, fever, malaise, drowsiness, and stupor.

Enterovirus meningitis: A cold, rash, or mouth sores with low-grade fever.

Poliomyelitis: *Initial*: nausea, vomiting, and cramps; *Active cases in CNS*: loss of motor function and severe paralysis.

Arbovirus encephalopathies: Sudden, high fever, severe headache, drowsiness, stiff neck; personality changes, seizures, and coma develop rapidly.

Herpes encephalitis: Acute onset of fever, headache, and change in mental status.

Rabies: *Initial*: headache, fever, abnormal sensations at bite site; *CNS*: periods of alert and aggressive behavior followed by muscle paralysis and hydrophobia.

Pathogenesis

Viral and viral-like diseases of the CNS are described below.

Aseptic Viral Meningitis

Poliomyelitis: A viral infection caused by the poliovirus (Picornaviridae). It is spread through poor hygiene or inhalation from infected individuals through pharyngeal secretions. After replication in the gastrointestinal tract, it can spread to the CNS. Nonparalytic polio leads to a **flaccid paralysis** in the lower limbs. Paralytic polio (myelitis) involves virus spread to the spinal cord, which can result in **respiratory paralysis** (**Figure 11A**).

Lymphocytic choriomeningitis (LCM): A mild, flu-like illness caused by the LCM virus (Arenaviridae). The virus infects the lymphocytes of the meninges.

Enterovirus meningitis: A common but rarely serious syndrome caused by the coxsackie viruses A and B or echoviruses (Picornaviruses). A GI-tract infection can spread to the meninges, leading to paralysis.

Viral Encephalitis

Herpes encephalitis: A rare infection of the temporal lobe caused by herpes simplex virus-1 (HSV-1; Herpesviridae). Following a latent infection, HSV spreads to the cranial nerves and the brain. Necrotic lesions in the temporal lobe lead to inflammation and encephalitis.

Rabies: A usually fatal, acute encephalopathy caused by the rabies virus (Rhabdoviridae). It is transmitted to humans in the saliva of a rabid animal bite. The virus slowly spreads through the blood to the CNS (**Figure 11B**). Once in the CNS, neuronal damage causes fatal encephalitis.

Arboviral Encephalopathies

An infected mosquito, carrying the virus from reservoir (horses and birds) to human, introduces the virus into the blood during a blood meal. **Primary encephalitis** occurs when the virus infects neurons in the brain, causing brain swelling and hemorrhage. Permanent nervous system damage may result. Several types of encephalitis caused by arboviruses occur in the United States.

St. Louis encephalitis (SLE): This form, caused by an arbovirus of the Flaviviridae, is the most common in the United States, especially in Texas and mid-western states.

La Crosse encephalitis: Caused by arboviruses of the Bunyaviridae, the disease mainly affects children in the mid-western and mid-Atlantic states.

Eastern equine encephalitis (EEE): This form, caused by an arbovirus in the Togaviridae, mainly affects young children and people over 55 years old along the eastern seaboard and Gulf Coast.

Western equine encephalitis (WEE): Another form caused by a virus in the Togaviridae, the disease can affect any individual but predominantly infants. The disease is typically found in the western United States.

Chikungunya fever: A form of encephalitis caused by the chikungunya virus (CHIKV; Togaviridae). CHIKV infection of the CNS can lead to encephalitis and causes debilitating joint pain or arthritis.

West Nile encephalitis (WNE): This virus is a member of the Flaviviridae and is closely related to the SLE virus. Several species of birds are the viral host. Encephalitis mainly affects the elderly. About 10% of infected individuals succumb to the disease.

Viral-Like Disease

Variant Creutzfeldt-Jacob disease (vCJD): A transmissible encephalopathy caused by a specific protein called a **prion** that affects the CNS. Most patients survive for 3–12 months after disease symptoms are diagnosed. Pneumonia is most often the cause of death.

Treatment

Treatment for viral meningitis involves bed rest, drinking plenty of fluids, and taking medications to relieve fever and headache. For rabies, the bite wound should be thoroughly cleaned with soap and water. The individual then should be given postexposure immunization (rabies immunoglobulin). For the arboviral encephalitides, treatment focuses on brain swelling, loss of the automatic breathing activity, and other treatable complications like bacterial pneumonia. Because antibiotics do not work on viral infections, they are not useful as a treatment option.

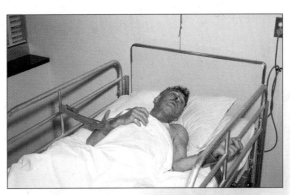

11A Polio. During the American polio epidemic of the 1950s, patients often needed the respiratory assistance of an "iron lung" to breath. (Courtesy of CDC)

11B A patient with rabies. Often the patient needs to be restrained because of sudden behavioral changes (e.g., agitation, terror, hallucinations) caused by the disease. (Courtesy of CDC)

12. Infections of the Heart and Bloodstream

Background

The **cardiovascular system** consists of the heart, blood, and blood vessels. When bacterial cells gain entry into the bloodstream, they can be transmitted to other parts of the body. **Bacteremia** refers to the presence of live bacteria in the blood. These conditions usually are not life threatening. Less common is **septicemia**, an infection where bacterial cells persist and divide in the blood. This condition is more serious than bacteremia and can lead to **sepsis**, which if not treated quickly, can be life threatening **septic shock**. In fact, about 33% of hospitalized patients with septic shock die.

By spreading through the blood, bacterial pathogens (often originating from a human reservoir) can find their way to the heart. Here, they may colonize the **endocardium**, the thin membranous lining of the heart muscle that also covers the heart valves and cause **endocarditis** (see right). Pathogens may also colonize the **myocardium**, the thick muscular wall of the heart; such an infection is called **myocarditis**.

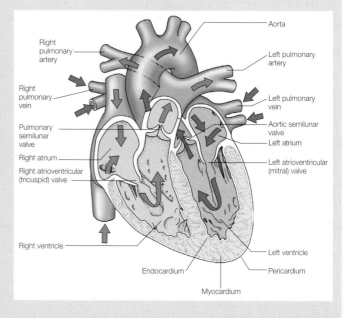

Right pulmonary artery
Right pulmonary vein
Pulmonary semilunar valve
Right atrium
Right atrioventricular (tricuspid) valve
Right ventricle
Aorta
Left pulmonary artery
Left pulmonary vein
Aortic semilunar valve
Left atrium
Left atrioventricular (mitral) valve
Left ventricle
Endocardium
Pericardium
Myocardium

 Signs and Symptoms

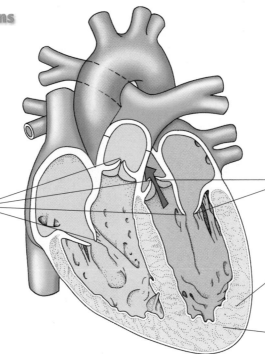

Sepsis: Fever, high heart and respiratory rates; *Severe sepsis:* decreased urine output, abrupt change in mental status, difficulty breathing, and abnormal heart function; *Septic shock:* severe sepsis plus low blood pressure.

Endocarditis: Fever, fatigue, fast heart rate, and an infection at another site.

Rheumatic heart disease: Complication from rheumatic fever; fever, joint pain, rash, chest pain, fatigue.

Myocarditis: Chest pain, fatigue, and abnormal heart rhythms.

American trypanosomiasis (Chagas disease): Fever, fatigue, rash, body aches, headache, loss of appetite, nausea, diarrhea or vomiting, swollen glands.

Pathogenesis

Bacterial diseases of the heart and bloodstream are described below.

Infections of the Endocardium

Infective (bacterial) endocarditis: A rare infection of the endocardium or heart valves involving bacteremia by a variety of bacterial species.

- **Acute endocarditis.** A sudden colonization of a damaged or diseased heart valve by a bloodstream infection involving *Staphylococcus aureus*, *Streptococcus pneumoniae*, *Streptococcus pyogenes*, or *Enterobacter faecalis*. Untreated, endocarditis is fatal as the valve leaflets fail to close completely leading to a backward flow of blood (**Figure 12A**).

- **Prosthetic endocarditis.** An acute or subacute onset of heart valve colonization due to *S. aureus* and some gram-negative bacilli. The prosthetic (artificial) valve often must be removed before antibiotic therapy can begin.

Rheumatic heart disease: A complication developing from rheumatic fever caused by *S. pyogenes* where antistreptococcal antibodies cross react with streptococcal antigens in the heart tissue. An inflammatory response can damage the heart valves, which can lead to permanent endocardial damage and heart failure (**Figure 12B**).

Infections of the Myocardium

Often a definite cause of myocarditis cannot be found. Viruses are common infectious agents, although worldwide, the most common cause is Chagas disease, an illness caused by a protozoan that is endemic to Central and South America.

Bacterial myocarditis: A rare form of myocarditis found mostly in immunocompromised patients.

Viral myocarditis: A rare form of myocarditis often caused by coxsackie B virus (CVB; Picornaviridae). The disease may be caused by direct cytopathic effects of virus, a pathologic immune response to persistent virus, or autoimmunity triggered by the viral infection. Most cases of CVB myocarditis resolve spontaneously, although in severe cases, the heart beat can weaken such that the heart cannot supply sufficient blood to the body.

American trypanosomiasis (Chagas disease): An acute but rare infection caused by the protozoal parasite *Trypanosome cruzi*. Following a bite by an infected reduviid bug, an acute phase disease involves the protozoal cells entering the bloodstream and lymphatics. Further infection of the cardiac muscle leads to myocarditis.

Septic Shock

Systemic inflammatory response syndrome (SIRS): An infection in the body (**sepsis**) that releases inflammatory mediators into the bloodstream. This can develop into a life-threatening syndrome if the uncontrolled inflammation leads to a sudden lowering of blood pressure and organ failure. This can trigger vascular collapse and **septic shock**.

Gram-negative sepsis: Often caused by *Escherichia coli*, *Enterobacter*, *Klebsiella pneumoniae*, and *Pseudomonas aeruginosa* when they release endotoxins (**endotoxin shock**).

Gram-positive sepsis: This, the most common form of sepsis, is caused by staphylococci and streptococci that produce exotoxins.

Treatment

Endocarditis can be treated with high doses of intravenous antibiotics for up to two months. Treatment of rheumatic fever includes antibiotics for the strep infection. Once the acute illness has been cleared, penicillin must be taken for many years to prevent recurrences and to lower the risk of heart valve damage should rheumatic fever recur.

Myocarditis treatment focuses on treating the underlying cause. If it is a bacterial cause, antibiotics may be prescribed. If viral, antiviral medications have not been effective in most cases. Treatment for acute phase Chagas disease involves using antiprotozoal drugs (benznidazole) to kill the parasite and manage signs and symptoms.

For severe sepsis or septic shock, broad spectrum antibiotics are given intravenously until the infectious agent is identified. Then, a narrow spectrum drug may be used that is effective against the identified bacterial species.

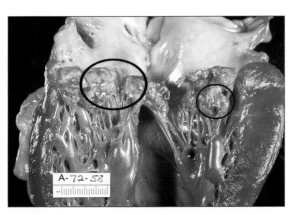

12A Bacterial endocarditis. A dissected left ventricle shows the mitral valve. Due to an infection with *Haemophilus parainfluenzae*, fibrin vegetations (ovals) have formed. (Courtesy of Dr. Edwin P. Ewing, Jr./CDC)

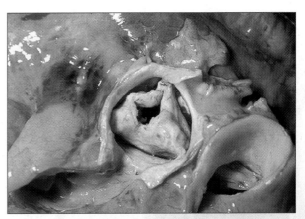

12B Rheumatic heart disease. With the aorta removed, the thickened, fused aortic valve leaflets can be seen. (Courtesy of Dr. Edwin P. Ewing, Jr./CDC)

13. Bacterial Infections

Background

As mentioned in the previous unit, the **cardiovascular system** consists of the heart, blood, and blood vessels. The **lymphatic system** is made up of the lymph, lymph vessels and nodes, and the lymphoid organs (e.g., spleen, tonsils, and appendix). Since fluids in the cardiovascular and lymphatic systems circulate throughout the body, the fluids come in contact with body tissues and organs. This means that many bacterial pathogens may spread systemically if they can gain entry to one or both of these systems.

The sources or **reservoirs** for these systemic bacterial infections and the mode of transmission are varied. Gas gangrene, for example, can develop in devitalized wounds that come in contact with soilborne endospores of *Clostridium perfringens*. Infections that come from animal reservoirs often are transmitted by direct contact. For example, brucellosis is found in grazing animals and is transmitted by contact with the infected animal. Likewise, tularemia is transmitted from infected rabbits to humans by contact (or incidentally by the bite of a deer fly or tick, making the latter case an example of vector transmission).

Other examples of animal reservoirs where arthropod vectors are the transmitting agent include plague (rodents and fleas), Lyme disease (deer mice and ticks), relapsing fever (rodents and soft ticks/lice), Rocky Mountain spotted fever (small mammals and ticks), typhus (rodents and lice/fleas), and ehrlichiosis and anaplasmosis (deer and ticks). Such diseases that are transmitted from wild or domestic animals to humans are called **zoonotic diseases**.

Bacterial infections from other animals also can involve transmission through a bite or scratch from the infected animal. Cat-scratch disease and rat-bite fever are examples.

According to the Centers for Disease Control and Prevention (CDC), about 75% of recently emerging human infectious diseases have an animal origin and some 60% of all human pathogens are zoonotic.

Signs and Symptoms

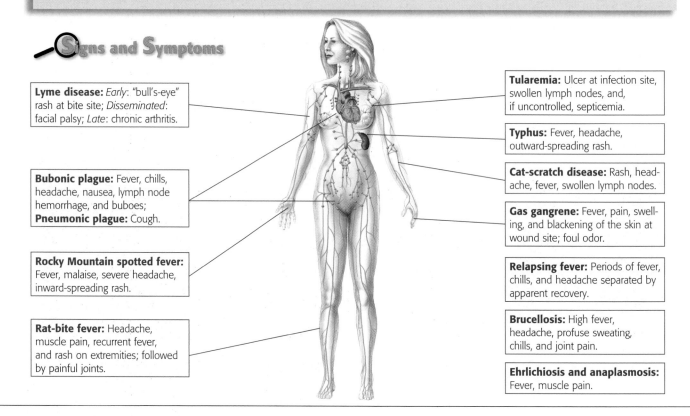

Lyme disease: *Early*: "bull's-eye" rash at bite site; *Disseminated*: facial palsy; *Late*: chronic arthritis.

Bubonic plague: Fever, chills, headache, nausea, lymph node hemorrhage, and buboes; **Pneumonic plague:** Cough.

Rocky Mountain spotted fever: Fever, malaise, severe headache, inward-spreading rash.

Rat-bite fever: Headache, muscle pain, recurrent fever, and rash on extremities; followed by painful joints.

Tularemia: Ulcer at infection site, swollen lymph nodes, and, if uncontrolled, septicemia.

Typhus: Fever, headache, outward-spreading rash.

Cat-scratch disease: Rash, headache, fever, swollen lymph nodes.

Gas gangrene: Fever, pain, swelling, and blackening of the skin at wound site; foul odor.

Relapsing fever: Periods of fever, chills, and headache separated by apparent recovery.

Brucellosis: High fever, headache, profuse sweating, chills, and joint pain.

Ehrlichiosis and anaplasmosis: Fever, muscle pain.

Pathogenesis

Systemic bacterial diseases are outlined below.

Contact Bacterial Diseases

Gas gangrene: A disease that occurs when devitalized wounds or surgical sites become infected with *Clostridium perfringens* endospores. Spore germination produces cells that infect the anaerobic wound and release toxins and enzymes causing extensive cell killing (**myonecrosis**) and promote systemic spread, where exotoxins can cause shock and renal failure.

Brucellosis (undulant fever): An acute or insidious disease often caused by contact with an animal infected with *Brucella* species. After spreading through the lymphatic system and into the blood, body organs (e.g., liver, spleen, kidneys) become infected. If untreated, patients may experience an intermittent (undulant) fever.

Tularemia (rabbit fever): A zoonotic disease transmitted by handling an animal (primarily rabbits) infected with *Francisella tularensis*. Besides causing an ulcer at the infection site, the pathogen also spreads to, and causes a swelling of, regional lymph nodes. Continued spread may lead to septicemia and death.

Vector-Transmitted Bacterial Diseases

Plague: A zoonotic disease caused by *Yersinia pestis*. Transmission usually occurs from infected rodents through the bite of an infected flea. The bloodborne bacilli most commonly enter the regional lymph nodes, leading to bubo formation (**bubonic plague**). Further blood dissemination (**septicemic plague**) to other body parts, including the lungs, causes pneumonia, which through aerosol transmission of the pathogen leads to **pneumonic plague** that untreated is almost 100% fatal.

Lyme disease: A zoonotic diseased resulting from the bite of an *Ixodes* tick infected with the spirochete *Borrelia burgdorferi*. The disease involves three stages: an early, localized stage (red, "bull's eye" rash; **Figure 13A**); a disseminated stage to the skin, heart, CNS, and joints; and a late stage in which chronic arthritis can develop.

Relapsing fever: A tick- or louse-borne disease caused by other species of *Borrelia*. In the bloodstream, new antigenic variants develop and reproduce, leading to recurring cycles of fever that become progressively shorter and milder.

Rocky Mountain spotted fever: A sudden onset disease transmitted by the bite of a wood or dog tick infected with *Rickettsia rickettsii*. Following an inflammation of the capillary endothelium, a rash develops on the palms and soles with an inward spread (**Figure 13B**). A few cases may cause renal and CNS damage leading to kidney and heart failure.

Typhus: Rickettsial diseases caused by the rupture of small blood (endothelial) vessels. The ensuing rash spreads outward from the trunk.

- **Epidemic typhus.** A potentially fatal form transmitted by lice infected with *Rickettsia prowazekii*. If untreated, endothelial invasion can lead to vascular necrosis (gangrene) on the hands and feet.

- **Murine (endemic) typhus.** The form transmitted by fleas infected with *Rickettsia typhi*. It is less severe than epidemic typhus.

Ehrlichiosis and anaplasmosis: Acute illnesses transmitted by ticks infected with *Ehrlichia chaffeensis* or *Anaplasma phagocytophilum*. Infection involves monocytes (ehrlichiosis) and neutrophils (anaplasmosis). Leukopenia can result in death.

Bacterial Diseases from Animal Bites

Cat-scratch disease: A fairly common, self-limiting infection caused by a cat bite or scratch that transmitts *Bartonella henselae* cells. Local lymph node swelling can occur.

Rat-bite fever: An abrupt onset infection caused by *Actinobacillus muris* (North America) or *Spirillum minus* (Asia). Pathogens in a rat bite cause a rash, and systemic spread can lead to endocarditis, pericarditis, or a brain infection in untreated late cases.

☤ Treatment

Early treatment of brucellosis and tularemia with antibiotics can be successful. Penicillin for gas gangrene will stop the spreading, but surgical removal of infected tissue may be required. Cat-scratch disease and rat-bite fever can be treated, if necessary, with antibiotics. Prompt treatment with streptomycin can decrease greatly the chance of death from bubonic plague. Likewise, early treatment with doxycycline is effective against Lyme disease and relapsing fever. Doxycycline also is effective against any of the rickettsial infections.

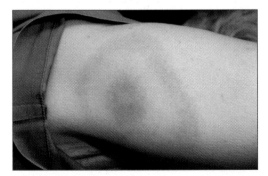

13A Lyme disease. The characteristic erythematous rash in the pattern of a "bull's-eye" at the site of the tick bite. (Courtesy of James Gathany/CDC)

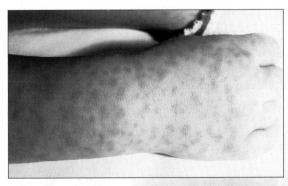

13B Rocky Mountain spotted fever. Infection produces a characteristic spotted rash that has extended inward from the palms. (Courtesy of CDC)

14. Viral and Parasitic Infections

Background

Several viruses and parasites can infect the cardiovascular and lymphatic systems. Protozoal diseases can have profound affects on humans, as malaria represents one of the most devastating of all human diseases, taking the lives of some 2 million people every year. Equally devastating are some helminthic diseases. Schistosomiasis infects 2 billion people, a third of the entire global population.

The **human immunodeficiency virus** (**HIV**), which is responsible for HIV disease and AIDS represents an immunodeficiency. Although the virus is often transmitted sexually, HIV does not attack the reproductive system. Rather, it eventually depletes **CD 4 (helper) T lymphocytes** that are essential in mounting an effective humoral and cell-mediated immune response. Without these cells, the immune system is almost helpless in fighting opportunistic infections arising from the HIV infection. Thus, a person usually does not die from the HIV infection but rather from secondary, **opportunistic infections** resulting from immune system destruction (AIDS).

The typical sequence of patient conditions leading to AIDS is:

1. **Primary HIV infection.** This is an acute HIV infection when anti-HIV antibody appears; flu-like symptoms may be present but soon disappear.
2. **Asymptomatic stage.** Although free from major symptoms, the patient has a slight decrease in T-cell count and mild leukopenia.
3. **Symptomatic stage.** The patient has a more severe drop in T cells and moderate leukopenia; weight loss, respiratory infections, and other opportunistic infections are the most common.
4. **AIDS.** An AIDS diagnosis is confirmed if a person with HIV exhibits severe opportunistic infections or specific cancers. Another indicator of AIDS is when the person's T helper cell percentage is below 14%, indicating serious immune damage has occurred.

Signs and Symptoms

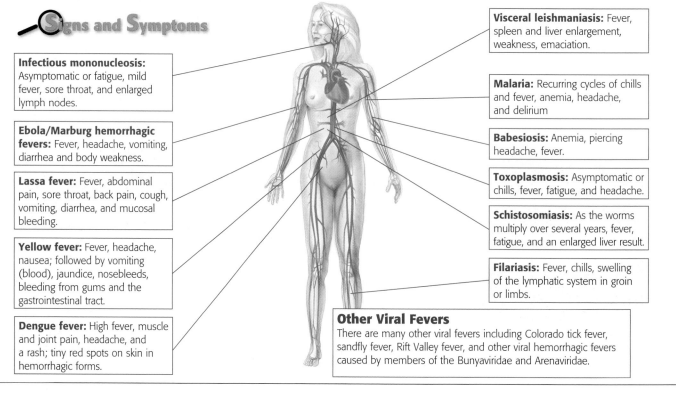

Infectious mononucleosis: Asymptomatic or fatigue, mild fever, sore throat, and enlarged lymph nodes.

Ebola/Marburg hemorrhagic fevers: Fever, headache, vomiting, diarrhea and body weakness.

Lassa fever: Fever, abdominal pain, sore throat, back pain, cough, vomiting, diarrhea, and mucosal bleeding.

Yellow fever: Fever, headache, nausea; followed by vomiting (blood), jaundice, nosebleeds, bleeding from gums and the gastrointestinal tract.

Dengue fever: High fever, muscle and joint pain, headache, and a rash; tiny red spots on skin in hemorrhagic forms.

Visceral leishmaniasis: Fever, spleen and liver enlargement, weakness, emaciation.

Malaria: Recurring cycles of chills and fever, anemia, headache, and delirium

Babesiosis: Anemia, piercing headache, fever.

Toxoplasmosis: Asymptomatic or chills, fever, fatigue, and headache.

Schistosomiasis: As the worms multiply over several years, fever, fatigue, and an enlarged liver result.

Filariasis: Fever, chills, swelling of the lymphatic system in groin or limbs.

Other Viral Fevers
There are many other viral fevers including Colorado tick fever, sandfly fever, Rift Valley fever, and other viral hemorrhagic fevers caused by members of the Bunyaviridae and Arenaviridae.

Pathogenesis

Viral and parasitic diseases of the cardiovascular and lymphatic systems are outlined below.

Viral Diseases

Infectious mononucleosis: A common viral syndrome resulting from direct ("kissing disease") or indirect contact with the Epstein-Barr virus (EBV; Herpesviridae). Oropharynx infection leads to infection of B lymphocytes resulting in an inflammation of lymph nodes and spleen. In immunocompromised individuals, B-cell infection can contribute to the development of **Burkitt's lymphoma**.

Viral Hemorrhagic Fevers

Yellow fever: A disease transmitted by mosquitoes infected with the yellow fever virus (Flaviviridae). In mild cases, viremia quickly resolves. Some cases progress to a liver infection, causing necrosis and hemorrhagic symptoms that can result in liver and kidney failure.

Dengue fever: Another acute disease transmitted by mosquitoes infected with the dengue fever virus (DFV; Flaviviridae). Lymphatic inflammation causes severe backache along with joint and muscle pain (**breakbone fever**). Should the individual become infected with a second DFV serotype, antibodies from the first infection cross react. Immune complex formation leads to hemorrhage and shock (**dengue hemorrhagic fever**).

Ebola and Marburg hemorrhagic fevers: Related hemorrhagic fever diseases transmitted by direct contact with an animal (e.g., fruit bats) infected with the virus (Filoviridae). Viremia leads to infection of numerous organs, causing necrosis and hemorrhaging. Shock and multiorgan failure result.

Lassa fever: An acute viral illness transmitted by direct or indirect contact with rodents infected with the Lassa fever virus (Arenaviridae). Viremia leads to a gradual onset of hemorrhaging similar to the Filoviridae.

Protozoal Diseases

Visceral leishmaniasis (Kala-azar): This, more serious form of the disease (see Unit 2), is transmitted by the bite of a sandfly infected with *Leishmania donovani*. Macrophage infection eventually takes the parasite to the spleen, liver, and bone marrow where it multiplies. Individuals with compromised immune systems can develop complications, which, due to secondary infections, often are fatal.

Toxoplasmosis: A blood and lymphatic system infection resulting from the ingestion of *Toxoplasma gondii* cysts in undercooked meat or cat feces. The invasive forms enter the bloodstream and become dormant in distant tissues. In pregnant women, the invasive forms cross the placenta and may cause birth defects.

Malaria: A disease transmitted by the bite of a mosquito infected with one of four species of *Plasmodium*. Once in the blood, the parasite infects the liver and then erythrocytes. Vessel blockage and hemorrhaging can occur. Organ damage can lead to a fatal drop in blood pressure.

Babesiosis: A mild disease (in healthy individuals) transmitted by ticks infected with *Babesia microti*. The parasite infects red blood cells and, in immunocompromised individuals, can cause a malaria-like illness.

Helminthic Diseases

Schistosomiasis (bilharzia): A blood-fluke infection transmitted by *Schistosoma* species. Skin penetration brings the fluke to the blood, where it can cause cirrhosis of the lungs and liver; egg release can produce an inflammation that blocks veins in the bladder, intestines, or liver (**Figure 14A**). A more mild form, **swimmer's itch**, only involves skin penetration and a localized inflammatory response.

Filariasis: A lymphatic system disease transmitted by the bite of a mosquito infected with *Wuchereria bancrofti*. Adult worms infect the lymph glands and vessels. Over years, the roundworms block lymph channels causing a swelling (**elephantiasis**) of the limbs or scrotum (**Figure 14B**).

☤ Treatment

No treatment usually is needed for infectious mononucleosis, and there is no effective treatment for dengue fever or the viral hemorrhagic fevers, although ribavirin can reduce the death rate from Lassa fever.

Stibogluconate is used for visceral leishmaniasis and sulfonamide drugs for toxoplasmosis. Malaria is treated with chloroquine and mefloquine. Schistosomiasis can be treated with praziquantal, while diethylcarbamazine is prescribed for filariasis.

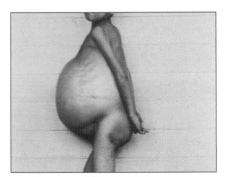

14A Schistosomiasis. An infection in the liver obstructs blood flow through the organ causing a buildup of fluid in the abdominal cavity. (Courtesy of WHO/TDR/Goldsmith & Kean. Used with permission.)

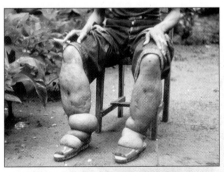

14B Filariasis. This parasitic infection, also called elephantiasis, blocks the lymphatic vessels causing fluid buildup in the legs. (Courtesy of CDC)

15. Infections of the Reproductive System

The reproductive system in males (below) and females (right) is involved with the production of gametes—sperm and eggs, respectively. These tracts normally are sterile except for the female vagina that is colonized by several types of resident microbiota. Several bacterial pathogens can infect the female reproductive tract, causing **vaginitis** (inflammation

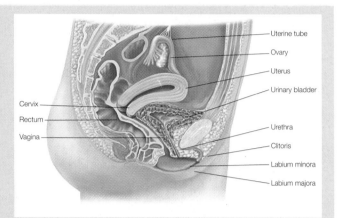

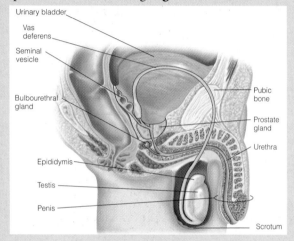

of the vagina), endometriosis (inflammation of the uterine lining), **salpingitis** (inflammation of the fallopian tubes), or **oophoritis** (inflammation of the ovaries).

Most pathogens cause **sexually transmitted diseases (STDs)**. Some STDs are systemic, while others remain associated with the female or male reproductive system. STDs also are referred to as **sexually transmitted infections** because often a person may be infected without displaying any symptoms.

Signs and Symptoms

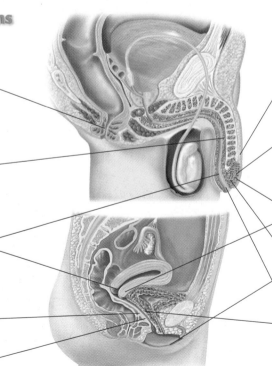

Lymphogranuloma venereum (LGV): Sore on genitals, fever, headache followed in 1–8 weeks by swollen lymph nodes that become buboes. Healing produces scars that can block lymphatic vessels.

Gonorrhea: *Male*: urethra inflammation, painful, burning urination, and pus discharge; *Female*: asymptomatic or burning urination, abdominal/pelvic pain, and pus discharge.

Chlamydial urethritis: Asymptomatic or painful urination, lower abdominal pain, vaginal/penile discharge; painful sexual intercourse (women); testicular pain (men).

Bacterial vaginosis: White vaginal discharge, painful urination, and vaginal itching.

Candidiasis: White vaginal discharge and vaginal itching.

Syphilis: *Primary*: chancre lasting 2–3 weeks; *Secondary*: skin rash, fatigue, fever, enlarged lymph nodes occurring 8–11 weeks after infection; *Tertiary*: neurological symptoms and gummas occur 5–40 years after infection.

Chancroid: Soft chancre on genitals and swollen lymph nodes.

Genital herpes: Fluid-filled vesicles appear that rupture into painful ulcers

Genital warts: Warts on external genitalia.

Trichomoniasis: *Female*: asymptomatic or itching, painful urination, and greenish-yellow discharge; *Male*: asymptomatic or urethritis and burning urination.

Pathogenesis

Reproductive system infections are summarized below.

Vaginitis

Bacterial vaginosis: A disease resulting from an overgrowth of the normal vaginal microbiota by *Gardnerella*, *Provotella*, and *Peptostreptococcus* species.

Candidiasis (vulvovaginitis): One of the most common causes of vaginitis that results from an infection by the fungus *Candida albicans*. Such "yeast infections" usually are opportunistic because of the loss of the resident microbiota through antibiotic use or immune suppression.

Sexually Transmitted Diseases—Bacterial

Syphilis: An infection with the spirochete *Treponema pallidum*. **Primary syphilis** characterized by a painless ulcerated lesion (**chancre**) at the contact site. Without treatment, the infection may spread (**secondary syphilis**), producing a red rash on the palms and soles. Many years later, the infection (**tertiary syphilis**) can involve damage to soft tissue and bone (**gummas**; **Figure 15A**). Blindness, insanity, and death can occur.

Gonorrhea: An acute infection caused by *Neisseria gonorrhoeae*. The pathogen infects the mucosal cells of the urethra and vagina causing an inflammation. In men, the infection leads to **gonococcal urethritis**. In women, the fallopian tubes can become infected (**gonococcal salpingitis**) and further infection of the uterus and/or ovaries leads to **pelvic inflammatory disease** (**PID**). Such infections run the risk of ectopic pregnancy.

Chlamydial urethritis: The most common STD in the United States. Caused by *Chlamydia trachomatis*, the pathogen attaches to mucous membranes of the urogenital tract and causes inflammation. If symptoms appear, men develop a **nongonococcal urethritis** (**NGU**) while women develop an infection of the cervix that may progress to PID.

Lymphogranuloma venereum (LGV): A distinct but rare infection caused by a different serovar of *C. trachomatis*. Painless ulcers form at the infection site and then heel as the pathogen spreads to the regional lymph nodes. Left untreated, it can cause proctitis or rectal blockage.

***Ureaplasma* urethritis:** Another form of nongonococcal urethritis caused by the mycoplasma *Ureaplasma urealyticum*, a common resident of the urogenital tract in men. It can result in male infertility and in women has been associated with salpingitis.

Chancroid: A disease caused by *Haemophilus ducreyi*. Infection produces a soft, painful chancre on the genitals. Swollen, painful inguinal lymph nodes develop in 50% of patients. Without treatment, chancre healing can damage or scar the genitals.

Sexually Transmitted Diseases—Viral and Protozoal

Genital herpes: A viral infection of the squamous epithelial cells usually caused by herpes simplex virus type 2 (HSV-2; Herpesviridae). Infection can be asymptomatic. In women, HSV-2 infects the cervix and vulva; in men, the penis (**Figure 15B**) or anus and rectum (through anal sex). Latent infections (in nerve ganglia) can be reactivated by stress and produce a milder vesicular infection at the primary infection site.

Genital warts: An infection primarily caused by human papilloma virus (HPV) 16 and 18 (Papovaviridae). Lesions are produced around the external genitalia. There is a high risk for development of cervical carcinoma.

Trichomoniasis: A common protozoal urogenital tract infection caused by *Trichomonas vaginalis*. In females, the pathogen causes vaginitis; in males, urethritis.

✚ Treatment

Metronidazole may be prescribed for bacterial vaginosis and an antifungal agent (miconazole or clotrimazole) for yeast infections.

Syphilis and gonorrhea are treated with penicillin, doxycycline is used for NGU, and tetracycline is effective on chancroid. PID also can be treated with cephalosporins and tetracycline. Oral acyclovir can be useful in treating primary infections of genital herpes. Metronidazole can be used for treatment of trichomoniasis.

15A Tertiary syphilis. The characteristic gumma on the nose of a patient with a long-standing syphilis infection. (Courtesy of J. Pledger/CDC)

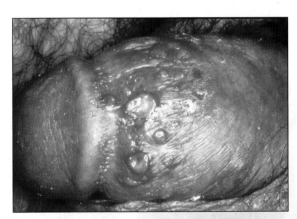

15B Genital herpes. A maculopapular herpetic rash has formed on the penile shaft. (Courtesy of Dr. N. J. Flumara and Dr. Gavin Hart/CDC)

16. Infections of the Urinary System

Background

The **urinary tract** consists of the kidneys, two ureters, the urinary bladder, and the urethra (see right). Because the outflow of urine from the bladder through the urethra is sterile, the urethra normally contains few resident microbes.

When infection does occur, such **urinary tract infections** (**UTIs**) may be **uncomplicated**, meaning there is no underlying condition that increases the risk of infection. Over 85% of UTIs are caused by bacteria from the intestine or vagina resulting in an ascending infection. **Complicated** UTIs involve catheters or another instruments that obstruct the tract.

Although about 50% of UTIs are asymptomatic, UTIs can exhibit several **clinical syndromes**:

Urethritis. An infection of the urethra.

Cystitis. An infection of the bladder usually resulting from a lower UTI infection involving enteric microbes.

Pyelonephritis. A bacterial infection of the kidneys.

Glomerulonephritis. An infection of the glomeruli of the kidneys resulting from an immune complex hypersensitivity.

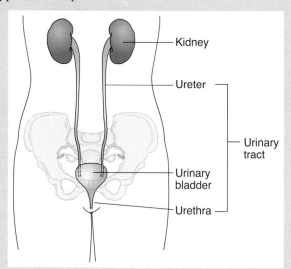

- Kidney
- Ureter
- Urinary tract
- Urinary bladder
- Urethra

Signs and Symptoms

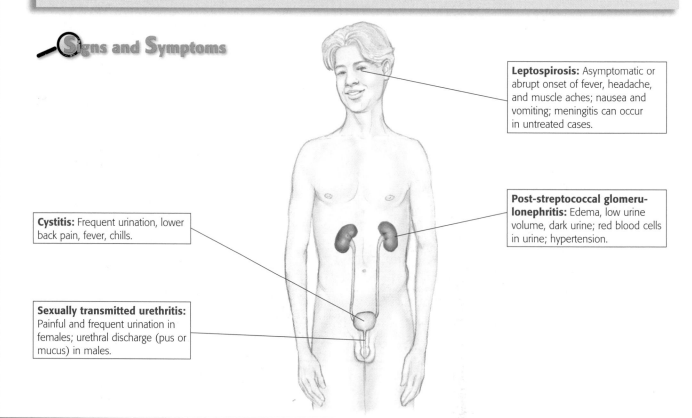

Leptospirosis: Asymptomatic or abrupt onset of fever, headache, and muscle aches; nausea and vomiting; meningitis can occur in untreated cases.

Post-streptococcal glomerulonephritis: Edema, low urine volume, dark urine; red blood cells in urine; hypertension.

Cystitis: Frequent urination, lower back pain, fever, chills.

Sexually transmitted urethritis: Painful and frequent urination in females; urethral discharge (pus or mucus) in males.